INTERDISCIPLINARY MATHEMATICS

BY ROBERT HERMANN

1. General Algebraic Ideas.
2. Linear and Tensor Algebra.
3. Algebraic Topics in Systems Theory.
4. Energy Momentum Tensors.
5. Topics in General Relativity.
6. Topics in the Mathematics of Quantum Mechanics.
7. Spinors, Clifford and Cayley Algebras.
8. Linear Systems Theory and Introductory Algebraic Geometry.
9. Geometric Structure of Systems -- Control-Theory and Physics, Part A.
10. Gauge Fields and Cartan-Ehresmann Connections, Part A.
11. Geometric Structure of Systems-Control Theory, Part B.
12. Geometric Theory of Nonlinear Differential Equations, Bäcklund Transformations, and Solitons, Part A.
13. Algebro-Geometric and Lie Theoretic Techniques in Systems Theory, Part A, by R. Hermann and C. Martin.
14. Geometric Theory of Nonlinear Differential Equations, Bäcklund Transformations, and Solitons, Part B.
15. Toda Lattices, Cosymplectic Manifolds, Bäcklund Transformations and Kinks, Part A.
16. Quantum and Fermion Differential Geometry, Part A.
17. Differential Geometry and the Calculus of Variations, 2nd Ed.
18. Toda Lattices, Cosymplectic Manifolds, Bäcklund Transformations and Kinks, Part B.
19. Yang-Mills, Kaluza-Klein, and the Einstein Program.
20. Cartanian Geometry, Nonlinear Waves, and Control Theory, Part A.
21. Cartanian Geometry, Nonlinear Waves, and Control Theory, Part B.
22. Topics in the Geometric Theory of Linear Systems.
23. Topics in the Geometric Theory of Integrable Mechanical Systems.

LIE GROUPS: HISTORY, FRONTIERS AND APPLICATIONS

Note: This series has outgrown its original plan, hence it will now bifurcate. SERIES A will continue the series of translations of the classics.

SERIES A

1. Sophus Lie's 1880 Transformation Group Paper. Translation by M. Ackerman, Comments by R. Hermann.
2. Ricci and Levi Civita's Tensor Analysis Paper, Translation and Comments by R. Hermann.
3. Sophus Lie's 1884 Differential Invariants Paper, Translation by M. Ackerman, Comments by R. Hermann.
4. Smooth Compactification of Locally Symmetric Varieties, by A. Ash, D. Mumford, M. Rapoport and Y. Tai.
5. Symplectic Geometry and Fourier Analysis, by N. Wallach.
6. The 1976 Ames Research Center (NASA) Conference on the Geometric Theory of Nonlinear Waves.
7. The 1976 Ames Research Center (NASA) Conference on Geometric Control Theory.
8. Hilbert's Invariant Theory Papers. Translation by M. Ackerman, Comments by R. Hermann.
9. Development of Mathematics in the 19th Century, by Felix Klein, Translated by M. Ackerman, Appendix "Kleinian Mathematics from an Advanced Standpoint", by R. Hermann.
10. Quantum Statistical Mechanics and Lie Group Harmonic Analysis, Part A, by N. Hurt and R. Hermann.
11. First Workshop on Grand Unification, by P. Frampton, S. Glashow, and A. Yildiz.
12. Inverse Scattering Papers: 1955-1963, by I. Kay and H.E. Moses.
13. Geometry of Riemannian Spaces, by Elie Cartan. Translated by J. Glazebrook, Commentary by R. Hermann.

SERIES B: SYSTEMS INFORMATION AND CONTROL

1. Geometry and Identification, P.E. Caines and R. Hermann, Eds.
2. Berkeley-Ames Conference on Nonlinear Problems in Control and Fluid Mechanics.

INTERDISCIPLINARY MATHEMATICS
VOLUME XXIII

TOPICS IN THE GEOMETRIC THEORY OF INTEGRABLE MECHANICAL SYSTEMS

ROBERT HERMANN

MATH SCI PRESS
53 JORDAN ROAD
BROOKLINE, MA 02146

Copyright © 1984 by Robert Hermann
All rights reserved

ISBN 0-915692-36-8

Library of Congress Cataloging in Publication Data

Hermann, Robert.
   Topics in the geometric theory of integrable mechanical systems.

   (Interdisciplinary mathematics ; v. 23)
   Includes bibliographies.
   1. Differential equations--Numerical solutions.
2. Hamiltonian systems. 3. Control theory. 4. Mechanics, Analytic. I. Title. II. Title: Integrable mechanical systems. III. Series.
QA371.H48 1984   515.3'5   84-15404
ISBN 0-915692-36-8

                    MATH SCI PRESS
                    53 JORDAN ROAD
                 BROOKLINE, MA 02146

Printed in the United States of America

TOPICS IN THE GEOMETRIC THEORY
OF INTEGRABLE MECHANICAL SYSTEMS

PREFACE

In Volume 22, I have collected my research work of the last five years on Linear System Theory. Similarly, this volume is concerned with material in that part of contemporary mathemical physics called the Theory of Integrable Systems. I have in mind development of relations between these mathematical parts of engineering and physics: I believe the connection might be most convincingly made via a notion of "approximability", which is yet to be made precise and workable. In fact, the idea of a relation between "approximability" and "integrability" was the foundation of the 19th century work, particularly that deriving from the Three Body Problem.

I see the theory of integrable systems, as it is developed in the contemporary mathematical physics literature, as a return to our roots in 19th century mathematics. Most of the great mathematicians of that period concerned themselves with one aspect or another of that topic. One can readily understand the practical motivation: The emphasis in both science and engineering was on working out the consequences of what we call Newtonian physics, and in "solving" differential equations, or, failing that, massaging them in such a way that useful information might be extracted. Galois, Abel, Liouville, and Lie tried to rationalize this process, and thereby started off much of the mathematics which has dominated our time!

Based on my own reading in 19th century mathematics, I would classify the attempts at creating a systematic rational theory of integrability of differential equations in two camps, which I will name after Lie and Weirstrass-Poincaré. The Lie camp is based on geometry and algebra, the Weirstrass-Poincaré on complex function theory.

Lie tried to systematize and classify differential equations in terms of their *symmetries* and *conservation laws*. In his study of ordinary differential equations he observed a correspondence between both. This has led to our modern theory of *symplectic manifolds*. Further, Lie's approach led to a partial classification of differential equations according to the symmetry groups associated with them. It is stretching things only a bit to see in our contemporary elementary particle physics a triumphant realization of Lie's point of view.

For the purpose of this book, let us think of this approach to integrability as an *algebraic* one. (Much of the work of Abel and Liouville on integrability

of differential equations is on the same point.) Another grand theme of the
19th century was the elaboration of the theory of elliptic and Abelian functions,
particularly the introduction of Theta Functions. Now, one can think of the
everywhere convergent power series of the Theta Functions as a sort of "integration" of the differential equations defining the elliptic functions. (They
are even quite efficient computationally.) Thus, a natural question for 19th
century mathematicians was:

> Which differential equations (ordinary or partial) admitted solutions
> as quotients of power series which converged in "large" regions,
> *independent of initial conditions*?

The Cauchy-Kowalewsky theorem gave an existence theorem for such
expressions *locally*: The idea was that certain equations were "integrable"
or "approximable" if such expressions could be found, preferably with explicit
recursive formulas for the coefficients. This is the theme that one can trace
through the work of Kowalewski and Painlevé for nonlinear ordinary differential
equations, and through Riemann and Fuchs and their successors for linear
ordinary differential equations.

I first encountered this material when I was a graduate student working
on the theory of Lie pseudogroups and associated geometric ideas and structures.
Parts II and III of Cartan's *Collected Works* (dealing with differential systems
and geometric structures, respectively) were a treasure-trove of ideas and
mathematics-to-be-developed. Part of our difficulty in understanding Cartan
was that we did not understand the mathematical milieu of the 1880's in which
he was reared: It was necessary to read and understand the work of some of
his predecessors and contemporaries (Darboux, Lie, Poincaré, Vessiot, Painlevé).
The theory of equivalence of geometric and differential equation structures
was central to the work of Lie and Cartan. In turn, this involved a sort of
Galois theory of differential equations which has yet to be developed, at
least in the geometric form envisaged by Lie and Cartan. All of this was too
much for me to digest or make very much of; to a certain extent, I have spent
the intervening years thinking about it. When it became important for
mathematical physics (especially the Korteweg-de Vries and Sine-Gordon work),
it became clear that these old ideas of "integrability" -- and associated
"approximatability" -- had finally come into their own.

In 1953 Charles Ehresmann was visiting Princeton and gave a course on his
theory of pseudogroups and prlongation of geometric structures. Although his
theory was very abstract and algebraic, it was based on a thorough knowledge of
classical differential geometry. Ehresmann himself was a student of Elie
Cartan, and had done important work on the topology of Grassmann manifolds.

Preface   v

   In his *Notice sur les Travaux Scientifiques*, written in this period, Ehresmann
says that, as a student at the Ecole Normale in the early 1930's, he had -- on
the advice of Vessiot -- studied Lie extensively. In the 1950's there was
great interest in elucidating Cartan's work on geometric structures and
infinite Lie groups. Don Spencer started here and pushed on to create his
masterpiece, the theory of Lie pseudogroups and their deformations. My own
inclination was to look for a more concrete manifestation of these very
general ideas. I was interested in the Lie theory of differential equations
and in "rationalizing" the diverse methods of solving explicitly low order
ordinary nonlinear differential equations. Ehresmann suggested that I look
at the work of Vessiot. When I did so, I came up to the great gulf between
the conceptual structure of geometry as it existed in the early 1900's and
the 1950's. (Curiously, it has changed very little since the 1950's, and
that mainly in the direction of returning to the more concrete and "classical"
concerns.) Not wanting to complicate my own entrance into research mathematics,
I put off the study of the Lie-Vessiot ideas and differential quations until
I had more leisure and knowledge, and, most important, had more concrete
problems as motivation. The interested reader will find bits and pieces of
this influence in my books and papers throughout the next twenty years.

   In 1975, I met Frank Estabrook and Hugo Wahlquist, who had introduced
interesting new ideas into the Korteweg-de Vries game, which had passed me by
in the 1960's when my applied interest was in the application of Lie group
theory to elementary particle physics. (I had caught something of the
greyhound mentality of the physicists I knew, and did not fully realize that
profound things could still be said about classical physics!) Formulating the
Estabrook-Wahlquist work, and the earlier work of Kruskal, Toda, et al., in my
geometric language (dealt with in previous volumes of *Interdisciplinary
Mathematics*) led me back to my earlier thoughts on Cartan, Lie, Vessiot, et al.
Of course, simultaneously, the mathematical physics world was enriched by a
new field, called *Integrable Systems*. Physicists and applied mathematicians
realized that more powerful algebraic and analytic technique enabled them to
go beyond the classical work and produce whole classes of such systems with
diverse and fascinating properties.

   When I was studying Lie and Vessiot in the 1950's, I also read the
closely related work of Painlevé, which had a different foundation, the
geometric theory of functions of one complex variable. (I had originally
thought of studying with Spencer because of my interest in that direction!)
I found that the second volume of Valiron's *Treaté d'Analyse* contained an
excellent exposition. (James Glazebrook is now preparing a translation
to be published by Math Sci Press.) This too has, of course, come roaring

back into fashion in mathematical physics in two different ways. Physically, the work of McCoy, Tracy and Wu has related it to the most important analytical theory in statistical mechanics, the Ising Model theory of Onsager. This has led to relations with the classical Riemann-Schlesinger theory of isomonodromy deformation of differential equations developed by M. Sato and his coworkers, the Chudnovsky brothers, as well as many others.

Since the theory of Integrable Systems as it has developed in the last twenty years came previously from the applied world, it has not developed and been expounded as it might have been if it had proceeded in the most logical and historical order. (Of course, this gives it excitement and vitality!) Since I was not the one who developed the ideas with the most originality, I have seen my role as that of intermediary between pure and applied worlds.

I have my own criticisms of both sides. The "applied" people have gloried in a certain conceptual anarchy. Since present-day geometry is hard for them to learn (because the "applied" motivation comes at the end of the process!), some of them have adopted a certain arrogance about things they really do not understand very well. ("Look Ma, I'm inventing Lie algebra!") On the other hand, the pure types often just see it as more grist for the big machinery, with their quota of theorems for the year. ("Look Ma, I'm doing Applied Mathematics!")

My goal in this book is to keep up with my self-appointed task of mixing things up: Call me an Applied Pure Mathematician. On the one hand, I am interested in the "big" machinery and geometric structure that I learned as a student in Princeton and absorbed in the other mathematical Centers which I have spent most of my mathematical life; but on the other, I am interested in discovering how that machinery must be modified to deal effectively with the problems which are continually being thrown up by science and technology. The point that the "pure" types do not seem to appreciate is that making these modifications and putting the theory effectively to work is a whole world in itself. Unless there is some encouragement for the process to take place, it will not happen, or will happen only on a time scale that is too long to do the maximal amount of benefit. I believe that the negative attitude of the pure mathematicians in the leading Centers is responsible for the fix the mathematical world finds itself in today. They listened to what Hilbert preached about doing Big Problems, not to what he actually *did*, i.e., talk to physicists and applied mathematicians.

In the area of Integrable Systems, I then see my role in adapting Lie, Vessiot, Painlevé, Cartan, Ehresmann, and Spencer, to the problems in the mathematical physics and applied mathematics world. I would also like to see these methods introduced into the control engineering world, but I have detected a certain resistance here, perhaps due to the ingrained mathematical conservation of those trained mathematically who work in engineering. They know quite well

Preface

what they learned as graduate students and its natural "analytic continuation", but are often astonishingly resistant to things off that beaten track. (This is in strong contrast to physicists, who will learn *anything* as soon as they smell its relevance to the "mainstream" problems which lead to the big glory.) Also, the sociology of the engineering world is not favorable: many senior engineering academics spend most of their time on what is basically money-raising, and in more-or-less subtle ways do not encourage their students to push on into new areas which they themselves do not understand.

There are many topics in mathematical engineering which seem ready for treatment with integrability ideas. The obvious one that seems to me to be more than ready, and to offer significant practical potential, is the theory of *stochastic systems*, and its associated theory of *filtering*. Certainly, the "Kalman-Bucy filter" is a beginning toward a more complete stochastic theory of integrability. The 1980 Les Arcs NATO Conference was held to pursue Lie-algebraic ideas about the possible generalization of the Kalman-Bucy filter, but they unfortunately did not lead anywhere. (although the conference did produce a magnificent *Proceedings*, which is our best current source of information about many *mathematical* aspects of filtering, identification, etc.) As Peter Caines has taught me, an important conceptual (and geometric) approach to the theory of stochastic systems seems to be in the "pure" treatises of Ikeda and Watanabe and Elworthy. They have a beautiful geometric formulation in terms of Levi-Civita parallel transport in Riemannian geometry. Tyrone Duncan has introduced these geometric ideas into the theory of filtering, and has also related stochastic systems to the theory of Kac-Moody Lie algebras. This work is in its early stages, but shows promise of leading to a theory of integrable stochastic systems, which must be there.

The basic analytical tool in the Kalman-Bucy filter is the matrix-Riccati equation. Clyde Martin and I have developed many of its properties in a very natural way by relating it to vector fields on Grassmann manifolds. (Of course, this was already essentially known to Lie!) It is also clear to one who knows the history that much of what Vessiot called the theory of Lie systems (which are the largest class of ordinary differential equations with some sort of reasonable finite-dimensional Galois theory attached to them) would come under a modern comprehensive theory of "Integrable Systems". This would realize the ancient idea that "integrability" is essentially a Galois-theoretic idea.

This volume itself is not even the beginning of a definitive treatment of the geometric theory of integrable systems that I (or a successor) will do some day. As my own thoughts gel, I write out material to develop them; it is more than just another paper, less, admittedly, than a polished book or review article that I might have written had I more time or a different temperament.

Although it plays no direct role here, I have in mind, for later development, my old interest, elementary particle physics. When I stumbled into this field in the 1960's, I found it difficult to comprehend because it lacked a rational geometric framework. I wrote my Benjamin books to give it such an impetus -- I wanted to use the self-evident importance of Lie groups and quantum field theory as a spur to development of certain mathematical ideas which might in turn reach back on the physics. In the 1970's the situation turned around completely -- I hope my books played a role -- and now avante-garde elementary particle physics seems completely dominated by geometric ideas. My thought is that the physical models which are now popular in elementary particle physics are, in a sense, on the boundary of what we think of now as Integrable Systems. For me, the great challenge has been to understand the geometric meaning of "renormalizability" of a field theory. It seems to me to have certain features of the Lie and Painlevé prototypes. Perhaps there is some class of systems extending the integrable case, which might be called "approximable"?

Of course, it would be most valuable to build a link between these "analytic" and "algebraic" themes. Here, the historical prototype was Sophie Kawolewski's work on this topic: Integrability of the equations of motions in the Weirstrass sense (i.e., as the quotient of everywhere convergent power series) led miraculously to integrability in the Lie sense! Conversely, "integrability" led (as Arnold has remarked) automatically to torii, hence to theta functions. The ramifications of this theme are yet far from completely understood! I envisage the creation of a theory of *integrable* and/or approximable control systems, deterministic and stochastic, based on alternate Lie algebraic and analytic foundations. Hence, the theory of approximation, as it is meant in numerical analysis and such hybrid topics as Padé and continued fraction expansions, should also find its natural engineering home!

Some of this work was supported by a grant from the Ames Research Center of NASA. I would like to thank Brian Doolin and George Meyer of Ames for encouraging me to pursue these byways as a side issue to my work in Geometric Control Theory, and Peter Caines for his helpful comments and insights on control mathematics and mechanics.

Karin Young continues her excellent editorial work and typing: I thank her again.

TABLE OF CONTENTS

	Page
PREFACE	iii

PART I: INTEGRABILITY, THE HAMILTON-JACOBI EQUATION, AND STACHEL THEORY ... 1

Chapter 1: INTEGRABILITY IN TERMS OF THE HAMILTON-JACOBI EQUATIONS ... 3

1. Introduction ... 3
2. Generation of Families of Solutions of the Hamilton *Ordinary* Differential Equations by a Single Solution of the Hamilton-Jacobi Equation ... 4
3. Some Hamiltonians Whose Ray Equations Take a Separate Form ... 6
4. Hamilton Systems of Liouville Type ... 8
5. The Liouville Seperability Condition and its General Meaning in Terms of Symplectic Manifold Theory and Cartan's Theory of Exterior Differential Systems ... 9
6. Planar Liouville Systems in Polar Coordinates ... 15
7. Hamiltonian Systems in Terms of "Nonholomonic" or Cartan Moving Frames ... 19
8. Liouville Systems in General Geometric Terms ... 22
9. Mechanical Systems Whose Configuration Space is SO(3,R) ... 23
10. Rigid Body Mechanics on the Nonholomonic Moving Frames ... 29
    References ... 31

Chapter 2: INTEGRABLE HAMILTONIAN SYSTEMS OF THE STÄCHEL TYPE ... 33

1. Introduction ... 33
2. The Stächel Strategy ... 34
3. The Stächel Conditions ... 36
4. The Liouville Conditions and Conformal Geometry ... 37
5. The Orbits of the Stächel Systems and their Quadratic Conservation Laws ... 37
6. The Stächel Infinitesimal Symmetries from the Lie Theoretic and Symplectic Point of View ... 41
7. On the Relation Between Integrability in the Sense of Liouville/Lie/Arnold and Algebraic Geometry ... 42
   References ... 46

PART II: GEOMETRIC STRUCTURES IN INTEGRABILITY THEORY ... 47

Chapter 3: THE CALOGERO EQUATIONS FOR INTEGRABLE SYSTEMS ... 49

1. Introduction ... 49
2. The Calogero Equations ... 50
3. The Calogero Equations as Curvature Conditions ... 51
4. The Calogero Equations in a More General Setting ... 54
   References ... 57

Table of Contents

	Page

Chapter 4: EHRESMANN PSEUDOGROUPS AND FOLIATIONS — 59

1. Ehresmann Pseudogroups on a Fixed Manifold — 59
2. Pseudogroups as the Automorphisms of Tensor Fields — 61
3. Examples of Locally Flat Tensor Fields and their Associated Pseudogroups Symplectic Structures — 62
4. Pseudogroups Generated by Lie Algebras of Vector Fields — 64
5. The Linearization of a Vector Field with a Fixed Point — 66
6. The Poincare Map Associated with a Periodic Orbit as the Linear Isotropy Subgroup of a One-Parameter Pseudogroup — 68
7. Pseudogroups and Holonomy Groups Associated with Foliations — 69
   References — 72

Chapter 5: LINEAR DIFFERENTIAL OPERATORS AND PARTIAL WAVE ANALYSIS — 75

1. Introduction — 75
2. Generalities about Lie Transformation Groups, Linear Differential Operators on Manifolds, and Orbit Spaces — 75
3. Differential Operators Invariant Under a One-Parameter Group — 77
4. Regular Lie Transformation Groups — 78
5. The Method of Fourier — 80

Chapter 6: LIE'S FUNCTION GROUPS AND POISSON STRUCTURES ON MANIFOLDS — 83

1. Introduction — 83
2. Poisson Operators on Manifolds and their Associated Tensor Fields — 83
3. The Schouten-Nijenhuis Tensor Associated with a Bivector Field — 85
4. Pseudogroups and Poisson Tensors — 87
5. Singular Foliations and the Frobenius Integrability Theorem: Implications for the Poisson Structure — 89
6. Homomorphisms of Bivector Fields and Function Groups in the Sense of Sophus Lie — 90
7. Homomorphisms of Poisson Structures and Function Groups in the Sense of Sophus Lie — 92
8. Lie "Function Groups" Generated by Lie Algebras of Vector Fields — 93
   References — 95

Chapter 7: PICARD-VESSIOT THEORY — 97

1. Introduction — 97
2. Differential Equations in Fiber Spaces, and Jet Sheaves of Solutions — 97
3. Differential Equations — 99
4. Prolongation — 99
5. Groups of Symmetries of Differential Equations and Symmetries in the Prolongation Sense — 101
6. Differential Operators and Differential Invariants and Covariants — 101
7. A General Framework for Picard-Vessiot
8. The Geometric Setting for the Classical Picard-Vessiot Theory — 102
9. The Classical Picard-Vessiot Theory — 107

Table of Contents   xi

                                                                                          Page

Chapter 8: RELATIONS BETWEEN KORTEWEG-DE VRIES AND PICARD-VESSIOT THEORY   111

    1. Introduction   111
    2. An Algebraic Setting. Lax Structures   111
    3. Some Examples of Lax Systems   113
    4. The Skew-Adjointness Condition for First Order Lax Systems   115
    5. The Muira Transform   117
    6. The Moyal Algebra   120
    7. The First Order Lax System on the Moyal Algebra   122

Chapter 9: THE GENERALIZED TODA LATTICES AS CAUCHY CHARACTERISTIC VECTOR FIELDS   125

    1. Introduction   125
    2. Cauchy Characteristics of Closed Two-Differential Forms   125
    3. Specialization to the Absolute Parallelism Defined by the Left Invariant Differential Forms on a Lie Group   128
    4. Flaschka Vector Fields on Vector Spaces   129
    5. The Flaschka Maps and Vector Fields for Reductive Lie Algebras   130
    6. Euler-Arnold Vector Fields on Lie Algebra   130
    7. Jacobi Triples of Lie Algebras   131
    8. Simple Root Systems for Simple Lie Algebras and Jacobi Triples   132
    9. Euler-Arnold Vector Fields that are Tangent to the Jacobi Subspaces   133
    10. Jacobi Triples Defined by Automorphisms of Lie Algebras   134
    11. Flaschka Maps Constructed from Jacobi Triples   135
    12. The "Integrability" Properties of Flaschka Vector Fields   136
        References   138

PART III: MECHANICS AND CONTROL   141

Chapter 10: TOWARD THE GEOMETRIC UNIFICATION OF OPTIMAL CONTROL THEORY, THE CALCULUS OF VARIATIONS, AND ANALYTICAL MECHANICS   143

    1. Introduction   143
    2. Notation and Some Basic Concepts of Differential Geometry   144
    3. The First and Second Variation in the Differential-Form Algebra   148
    4. The First and Second Variation for "Canonical" Variational Problems   150
    5. The Second Variation for Variational Problems in Canonical Form   153
    6. Extremal Fields in the Differential-Form Variational Formalism   155
    7. Newton's Laws and Characteristics of Two-Differential Forms   156
    8. The One-Jet Space and its Contact Forms   157
    9. The Statement of Newton's Laws in Terms of the One-Jet Space   158
    10. Mechanical Systems which have the Same Newton-Lagrange Vector Field   159
    11. Symplectic Structures on the Space of all Trajectories of a Regular Mechanical System   160
    12. When is the Two-Differential Form Defined by the Newton-Lagrange Equations Closed?   162
    13. The Optimal Control Problem in Canonical Form   166
    14. Maxwell's Equations   170
        References   173

Page

Chapter 11: MAXWELL'S ELECTROMAGNETIC EQUATIONS AND ANALYTICAL MECHANICS

1. Introduction — 177
2. The Lagrangian for Maxwell Theory — 177
3. First Variation of the Lagrangian and Maxwell's Equations — 178
4. The Algebra of Maxwell's Equations — 178
5. Kinetic and Potential Energy in Newtonian Mechanics — 181
6. Kinetic and Potential Energy in a Relativistic Form — 181
7. Kinetic and Potential Energy for Maxwell Fields — 182
8. Force in Newtonian Particle Mechanics in Terms of the Calculus of Variations — 184

Chapter 12: PERIODIC SOLUTIONS FOR THE MATRIX RICCATI EQUATION VIA LIE THEORY — 189

1. Introduction — 189
2. Certain Types of Periodic Orbits as Fixed Points — 190
3. A Lie-Theoretic Situation — 192
4. The Case where G/H is a Coset Space such that (G,H) is a Grassmann Pair — 193
5. Periodic Orbits of Certain Vector Fields on Compact Homogeneous Spaces — 195
6. Classes of Vector Fields on the Lagrange-Grassmann Manifold that have Periodic Orbits — 196

PART IV: STOCHASTIC SYSTEMS

Chapter 13: DIFFERENTIAL GEOMETRY AND LIE THEORY OF CLASSICAL AND QUANTUM STOCHASTIC SYSTEMS — 203

1. Introduction — 203
2. Diffusion Equations on Manifolds — 205
3. Quantum Diffusion Operators — 207
4. The Algebra of the Quantum Mechanical FP Operators — 208
5. Some Classical and Quantum Operators of FP-Type — 210
6. Infeld-Hull Factorization and Solution of Linear Evolution Equations — 213
7. Factoring Second Order Differential Operators in Commutative Differential Algebras — 215
8. Differential Equations whose Lie Algebras are Finite Dimensional — 217
9. Generalization of the Fock-Bargmann-Segal Construction — 218
10. An Abstract Form of Picard-Vessiot Theory — 221
11. Factoring Second Order Differential Operators with Rational Coefficients — 223
12. The Infeld-Hull Factorization of the Bessel Equation — 226
13. The Infeld-Hull Factorization of the Bessel Equation in Terms of Deformation Theory — 228
14. The Infeld-Hull Factorization for the Whittaker Equation — 229
15. The Lie Algebra and Infeld-Hull Structure for the Legendre Function — 230
16. The Lie Algebra of the Infeld-Hull Relations in Terms of the Rational Enveloping Algebras of Finite Dimensional Lie Algebras — 231
17. Quantum Stochastic Systems whose Deterministic Part is the Harmonic Oscillator in the Fock-Segal Representation — 233

References — 237

Table of Contents   xiii

                                                                            Page

Chapter 14: LIE THEORY AND STOCHASTIC SYSTEMS, PART II. THE
             GEOMETRY OF A PROBABILISTIC LIE THEORY   239

   1. Introduction   239
   2. Tangent Vectors and Vector Fields for the Space of
      Probability Measures   239
   3. General Fokker-Planck Operator   242
   4. The Lie Structure of F-P Operators   242
   5. The Fokker-Planck Operators on Filtered Lie Algebras   244
   6. Lie Algebras of Second-Degree, One-Dimensional Fokker-Planck
      Operators   245
   7. Lie Algebra of Second Degree Fekker-Planck Operators
      Determined by the Conformal Structure of a Riemannian Manifold   246
   8. Quantum Mechanics and Cotangent Bundle of the Space of
      Probability Measures   247
   9. Estabrook and Harrison's Treatment of the Symmetries of the
      One-Dimensional Heat Equation Using the Theory of Exterior
      Differential Systems   249

PART V: KRON-KONDO THEORY

Chapter 15: THE METHODS OF KRON, HOFFMAN AND KONDO IN GEOMETRIC
             SYSTEM THEORY   257

   1. Introduction   257
   2. Connections in Vector Bundles   258
   3. Generalized Linear Input-Output Systems and Covariant
      Derivatives. Left Invariant Mechanics on Lie Groups   261
   4. Linear Systems Coupled to Geometric Structures   262
      References   262

Chapter 16: QUASI-COORDINATES AND MOVING FRAMES FOR LAGRANGIAN
             MECHANICAL SYSTEMS   263

   1. Introduction   263
   2. Lagrangian Mechanical Systems on Manifolds   263
   3. Quasi-Coordinates and Bases of Differential Forms/Moving
      Frames   265

Chapter 17: CYCLIC COORDINATES AND LINEAR SYSTEMS   269

   1. Introduction   269
   2. Symmetries of Mechanical Systems   269
   3. Quotient Maps for Mechanical Systems with Groups of Symmetries   270
   4. Cyclic Coordinates in the Classical Sense for Hamiltonian
      Systems   271
   5. Completely Degenerate Lagrangians   273
   6. Cyclicity from the Point of View of Lagrange's Equations   276
   7. Systems with Two Degrees of Freedom whose Configuration
      Coordinates are Cyclic   277
   8. Linear Cyclicity   279
   9. The Lagrange-Rayleigh Equations with External Forces   282
  10. Direct Product of Lagrange-Raleigh Systems with External
      Forces   284
  11. Lagrange-Rayleigh Systems with Non-Holonomic Constraints   285
  12. Linear Lagrange-Raleigh Systems with Kirkhoffian Constraints   287

	Page

Chapter 18: DIFFERENTIAL GEOMETRY OF ENGINEERING-MECHANICS SYSTEMS — 291

1. Introduction — 291
2. The Lagrange Equations — 292
3. Lagrangian Mechanics with Constraints — 294
4. Input-Output Theory for Mechanical Systems — 295
5. Interaction of Lagrangian Systems — 296
6. Condition that Lagrange's Equations be Linear, Time-Invariant — 298
7. Quadratic Lagrangians and Linear Equations — 298
8. F = MA and the Ehresmann Jet-Calculus — 299
9. A General Definition of Mechanical Systems in Terms of Exterior Differential Systems — 301
10. The Two-Jet Space as a Vector Bundle over the One-Jet Bundle — 304

Chapter 19: NEWTON-LAGRANGE LINEAR SYSTEMS — 309

1. Introduction — 309
2. Quadratic Lagrangian — 310
3. Linear Newton-Lagrange Systems for which q is a Cyclic Vector — 312

Chapter 20: THE GEOMETRIC NATURE OF "POWER" IN LAGRANGIAN MECHANICS ON MANIFOLDS — 315

1. Introduction — 315
2. Lagrangian Systems — 315
3. Lagrange's Equations with Forces and Control — 316
4. Energy and Power — 317
5. The Space of One-Jets — 318

Chapter 21: THE RLC EQUATION IN GEOMETRIC FORM — 323

1. Introduction — 323
2. The Damped-Harmonic Oscillator-RLC Equations in Coordinate-Free Frames — 323
3. Power — 324
4. The "Transfer Function" of the RLC-System as a Rational Map from $P_1(\mathbb{C})$ to a Grassmannian — 324
5. State Space Form for RLC Systems in the Non-Singular Case — 325
6. RLC Equations Coupled to Riemannian Metrics -- The Kron-Hoffman-Kondo Equations — 327

PART VI: GEOMETRIC STRUCTURE IN FIELD THEORY — 331

Chapter 22: THE HAMILTON-VOLTERRA FIELD THEORY EQUATIONS AND THE COJET BUNDLES — 333

1. Introduction — 333
2. The Volterra Formalism in Local Coordinates — 337
3. The Hamilton-Volterra Equation in Terms of Differential Forms and Exterior Differential Systems — 338
4. The Hamilton-Volterra Equations and the One-Cojet Bundle — 340
5. Volterra Tensors — 343
6. The Natural Class of Field Theories Containing the Harmonic Maps of Eels and Sampson, the σ-Model of the Elementary Particle Physicists, and Generalizing the "Newtonian" Mechanical Models — 345
   References — 346

PART I

INTEGRABILITY, THE HAMILTON-JACOBI
EQUATION, AND STACHEL THEORY

Chapter 1

INTEGRABILITY IN TERMS OF THE
HAMILTON-JACOBI EQUATIONS

1. INTRODUCTION

In the classical literature, there are basically two approaches to integrable mechanical systems: One based on symmetries, which I will call the *Lie method*, the other based on the existence of *separable* solutions of the associated Hamilton-Jacobi partial differential equations. Both methods are described in one form or another in the classical treatises. As usual, Whittaker [1] is best for its comprehensiveness and conceptual directness and concreteness, although it requires long effort and study to translate its material into a form understandable and congenial to a modern reader.

In this chapter, I will concentrate on the geometry of the Hamilton-Jacobi equations. Here, the best classical references I have found to be the treatises by Gantmacher [2] and Pars [3]. What is particularly interesting about this classical work is that it has close links with classical algebraic geometry, particularly the theory of Abelian Integrals on Riemann surfaces. In fact, more recent work on integrable systems flowing out of the Korteweg de Vries and Toda Lattice equations also involves this mathematics. The reasons are related (and to a certain extent "explained" by a general argument by Arnold [4]), but I think it is of value to study the older methods from an independent point of view. What is also of interest is that the classical methods seem to be oriented in a Riemannian geometric and connection-theoretic direction.

My own interest in this classical work is that it seems to involve interesting material on *deformation theory*, particularly the theory of deformation of the differential equations, for reasons I will explain below. See Ref. [5] for background in mathematics and the system theoretic applications I have in mind. A survey paper I am writing for the *Proceedings of the 1983 APSM AMES-Berkeley Workshop on Nonlinear Geometric Methods in Aircraft Control and Fluid Mechanics* (to be published by Math Sci Press) will go into this in more detail.

I will start off by reviewing material from Hamilton-Jacobi theory, in the classical coordinate notation. This material can be readily translated into the modern coordinate-free symplectic manifold approaches. See my *Differential Geometry and the Calculus of Variations* (Interdisciplinary Mathematics, Volume 17) and Abraham and Marsden's *Foundations of Mathematics*.

A new geometric feature to my treatment is that I emphasize the correspondence between a complete solution of the Hamilton-Jacobi partial differential equation and the *general solution* of the Hamilton equations, rather than, as in the Jacobi method, the canonical transformation to *action-angle* variables. It seems to me that my method is better adapted to geometric clarity and precision, as well as lending itself better to possible generalization and application.

2. GENERALIZATION OF FAMILIES OF SOLUTIONS OF THE HAMILTON *ORDINARY* DIFFERENTIAL EQUATIONS BY A SINGLE SOLUTION OF THE HAMILTON-JACOBI EQUATION

Let X be a $2n$ dimensional manifold, with a symplectic structure defined by a closed two-form ω of maximal rank. A *coordinate system*

$$(p_i, q^i), \quad 1 \leq i, j \leq n$$

in an open subset U of X is said to be *canonical* if

$$\omega = dp_i \wedge dq^i \tag{2.1}$$

(Summation convention in force.) For the rest of this section we will work, in the classical style, in this fixed coordinate system.

Let

$$H: X \to R$$

$$(q,p) \to H(q,p)$$

be a real-valued function. *Hamilton's equations* are the following ordinary differential equations:

$$\frac{dq^i}{dt} = \frac{\partial H}{\partial p_i}(q(t), p(t))$$

$$\frac{dp_i}{dt} = -\frac{\partial H}{\partial q^i}(q(t), p(t)) \quad . \tag{2.2}$$

The *Hamilton-Jacobi equation* is the following partial differential equation to be solved for a map $(q,t) \to S(q,t)$:

$$\frac{\partial S}{\partial t} + H\left(q, \frac{\partial S}{\partial q}\right) = 0 \tag{2.3}$$

<u>Theorem 2.1</u>. For each solution $(q,t) \to S(q,t)$ of (2.3), each of the solutions $t \to (q(t), p(t))$ of the following equations is a solution of (2.2):

$$p_i(t) = \frac{\partial S}{\partial q^i}(q(t),t) \qquad (2.4)$$

$$\frac{dq^i}{dt} = \frac{\partial H}{\partial p_i}\left(q(t), \frac{\partial S}{\partial q^i}(q(t),t)\right) \qquad (2.5)$$

Proof. (2.4) and (2.5) together imply the first set of the Hamilton equations (2.2). We need only verify that, with a solution $t \to q(t)$ of (2.5), and $t \to p(t)$ *defined* by (2.4), the curve

$$t \to (q(t), p(t))$$

on X satisfies the set of equations in (2.2).

Differentiate (2.4):

$$\frac{dp_i}{dt} = \frac{\partial^2 S}{\partial q^i \partial q^j} \frac{dq^j}{dt} + \frac{\partial^2 S}{\partial t \partial q^i}$$

$$= \text{, using (2.5) and (2.3),}$$

$$\frac{\partial^2 S}{\partial q^i \partial q^j} \frac{\partial H}{\partial p^j} - \frac{\partial}{\partial q^i}\left(H\left(q, \frac{\partial S}{\partial q}\right)\right)$$

$$= -\frac{\partial H}{\partial q^i}$$

as required.

Remark. This simple result is the analytical underpinning to both the *method of extremal fields* in the Calculus of Variations and Geometric Optics. It also fits in very well with Cartan's approach via the coordinate-free theory of exterior differential systems. *Differential Geometry and the Calculus of Variations* contains more details.

Thus, to each *single* solution S of the Hamilton-Jacobi partial differential equation, we assign an n-parameter family of solutions of the Hamilton ordinary differential equation. If we have an n-parameter family

$$(q,t) \to S(q,t; a_1, \ldots, a_n)$$

of Hamilton-Jacobi functions satisfying the nondegeneracy condition

$$\det\left(\frac{\partial^2 S}{\partial q^i \partial a_j}\right) \neq 0 \qquad (2.6)$$

we obtain a 2n-parameter family of solutions of the Hamilton equation:

$$t \to q(t;q_0,a)$$

$$\frac{\partial q^i}{\partial t}(t;q_0;a) = \frac{\partial H}{\partial q^i}\left(q, \frac{\partial S}{\partial q}\right) \qquad (2.7)$$

$$q^i(0;q_0,a) = q_0^i .$$

Theorem 2.2. The 2n-parameter family of solutions of the Hamilton ordinary differential equation is a *general solution*, in the sense defined in Interdisciplinary Mathematics, Volume 12 or Ref. [7].

Proof. Left to the reader. It can also be proved directly as a consequence of Jacobi's canonical transformation to "action-angle variables".

Our strategy is to now look for conditions on H which guarantee that the Equations (2.7) are "integrable by quadratures" in the classical sense. The simplest way of doing this is to suppose that they are *separable*, in the sense that they reduce to the following form:

$$\begin{aligned}\frac{\partial q^1}{\partial t} &= f^1(q^1;a_1) \\ &\vdots \\ \frac{\partial q^n}{\partial t} &= f^n(q^n;a_n)\end{aligned} \qquad (2.8)$$

In any case, we see that the subject is a marvelous testing ground for a *deformation theory of differential equations degrading on parameters*, such as I am developing in Ref. [5].

3. SOME HAMILTONIANS WHOSE RAY EQUATIONS TAKE A SEPARATE FORM

Continue with (q,p), $H(q,p)$, as above. If $(q,t) \to S(q,t)$ is a solution of

$$\frac{\partial S}{\partial t} + H\left(q, \frac{\partial S}{\partial q}\right) = 0 \qquad (3.1)$$

we will call the following ordinary differential equations the *ray equations* associated with the single solution S of (3.1)

Hamilton-Jacobi

$$\frac{dq}{dt} = \frac{\partial H}{\partial p}\left(q, \frac{\partial S}{\partial q}\right) \tag{3.2}$$

From now on, we will consider only solutions of (3.1) of the following form:

$$S(q,t) = Et + W(q) , \tag{3.3}$$

where $q \to W(q)$ is a scalar valued function of q *alone*, and E is a *constant*. (Physically, it is *total energy*, which is the reason for the notation.) The ray equations then take the following form:

$$\frac{dq}{dt} = \frac{\partial H}{\partial p}\left(q, \frac{\partial W}{\partial q}\right) \tag{3.4}$$

For each value of the constant E, let

$$X_E = \{(q,p): H(q,p) = E\} \tag{3.5}$$

X_E is the *energy surface*.

Definition. S of form (3.3) is said to be a *separable* solution of the Hamilton-Jacobi equation if the following conditions are satisfied:

There are functions

$$W_1(\ ), \ldots, W_n(\ )$$

of *one* variable such that

$$W(q^1, \ldots, q^n) = W_1(q^1) + \cdots + W_n(q^n) \tag{3.6}$$

There are n functions

$$H^i(\ )$$

of *three* real variables such that:

$$\frac{\partial H}{\partial p^1}\left(q, \frac{\partial W}{\partial q}\right) = H^1\left(q^1, \frac{dW^1}{dq^1}, E\right)$$

$$\vdots \tag{3.7}$$

$$\frac{\partial H}{\partial p^n}\left(q, \frac{\partial W}{\partial q}\right) = H^n\left(q^n, \frac{dW^n}{dq^n}, E\right)$$

Of course, conditions (3.7) are precisely those which guarantee that the ray equations are of separable type, and have to be solved "by quadratures"

$$\frac{dq^1}{dt} = H^1\left(q^1, \frac{dW^1}{dq^n}, E\right)$$
$$\vdots$$
$$\frac{dq^n}{dt} = H^n\left(q^n, \frac{dW^n}{dq^n}, E\right) \qquad (3.8)$$

I will not enter here into further necessary and sufficient conditions for separability. This is covered by Stachel's classical work. See the treatment in Pars' treatise [3].

We can of course "solve" equations (3.8) by quadratures:

$$\int \frac{dq^1}{H^1\left(q^1, \frac{dW^1}{dq^1}; E\right)} = t$$
$$\vdots \qquad (3.9)$$
$$\int \frac{dq^n}{H^n\left(q^n, \frac{dW^n}{dq^n}; E\right)} = t$$

Often, in practice, the denominators in the integrands of (3.9) will be, as functions of q, *rational* functions on an algebraic curve

$$F(q,z) = 0 \qquad (3.10)$$

Thus, actually solving the equations (3.9) (hence, finding an n-parameter family of solutions of the Hamilton equations) will involve "inversion" of the "Abelian integrals" (3.9) on the algebraic curve (3.10), which is *the* classical problem of algebraic geometry!

Let us now turn to the Liouville system, which provides an extensive class of equations separated in the above sense.

4. HAMILTON SYSTEMS OF LIOUVILLE TYPE

Keep the above notation.

<u>Definition</u>. H is said to be of *Liouville type* if it can be written in the following form:

$$H(q,p) = \frac{H_1(q^1_1, p_1) + \cdots + H_n(q^n_n, p_n)}{A_1(q^1_1, p_1) + \cdots + A_n(q^n_n, p_n)} \qquad (4.1)$$

Theorem 4.1. If H is of the Liouville type, then it admits (locally) an n-parameter family of solutions of the Hamilton-Jacobi equation of the following form:

$$S(q,t) = Et + W_1(q_1, E_1) + \cdots + W_n(q_n, E_n) \qquad (4.2)$$

with

$$E_1 + \cdots + E_n = 0 . \qquad (4.3)$$

Proof. Chose the $W_1, \ldots, W_n$ so that they are solutions of the following *ordinary* differential equations. Then there are constants $E_1, \ldots, E_n$ such that

$$\begin{aligned}
H_1\left(q^1, \frac{dW_1}{dq^1}\right) - EA_1\left(q^1, \frac{dW_1}{dq^1}\right) &= E_1 \\
H_2\left(q^2, \frac{dW_2}{dq^2}\right) - EA_2\left(q^2, \frac{dW_2}{dq^2}\right) &= E_2 \\
&\vdots \\
H_n\left(q^n, \frac{dW_n}{dq^n}\right) - EA_n\left(q^n, \frac{dW_n}{dq^n}\right) &= E_n
\end{aligned} \qquad (4.4)$$

Adding up both sides of (4.4), using (4.9), shows that S defined by (4.2) solves the Hamilton-Jacobi equation if the Hamiltonian H is of Liouville form (4.1).

5. THE LIOUVILLE SEPERABILITY CONDITION AND ITS GENERAL MEANING IN TERMS OF SYMPLECTIC MANIFOLD THEORY AND CARTAN'S THEORY OF EXTERIOR DIFFERENTIAL SYSTEMS

Now let us investigate the meaning of the Liouville condition from the viewpoint of coordinate-free differential geometry.

Let X be a manifold of dimension $2n$, ω a closed two-form on X which defines a symplectic structure, and

$$H: X \to R$$

a C^∞ real-valued function. Then, there is a unique vector field V_H such that:

$$H = -V_H \, \lrcorner \, \omega \qquad (5.1)$$

The orbits of V_H are, in local coordinates (q^i, p_i) such that $\omega = dp_i \wedge dq^i$, i.e., ω takes its *canonical form*, the solutions of *Hamilton's equations* with *Hamiltonian* H.

<u>Definition</u>. A *Hamilton-Jacobi submanifold* of the structure (M, ω, H) is an n-dimensional submanifold

$$\phi: N \to M$$

satisfying the following condition:

a) $\quad \phi^*(H) = -E$, constant $\hfill (5.2)$

b) $\quad \phi^*(\omega) = 0 \hfill (5.3)$

<u>Example</u>. Suppose $M = R^{2n}$ with canonical coordinates (p_i, q^i), i.e.,

$$\omega = dp_i \wedge dq^i$$

Suppose that

$$N = R^n$$

and

$$\phi: N \to M$$

is of the following form:

$$\phi^*(p^i) = \frac{\partial W}{\partial q^i}$$

$$\phi^*(q^i) = q^i \quad , \hfill (5.4)$$

for a function $W(q)$ on M. Then

$$\phi^*(p_i, q^i) = \frac{\partial W}{\partial q^i} dq^i$$

$$= dW \hfill (5.5)$$

hence, (5.3) is satisfied. That (5.2) is satisfied is *precisely* the condition that

$$q \to W(q)$$

satisfies the Hamilton-Jacobi equation

Hamilton-Jacobi

$$H\left(q, \frac{\partial S}{\partial q}\right) = -E .$$

Thus, the vector field on Q:

$$R_W = \frac{\partial S}{\partial q^i} \frac{\partial}{\partial q^i} \qquad (5.6)$$

has the following property:

$$\phi_*(R_W) = V_H \quad \text{restricted to} \quad \phi(N) \qquad (5.7)$$

which mean, analytically, that the orbit curves of R_W are, when mapped by ϕ_S, solutions of the Hamilton equations with Hamiltonian H. R_W is the field of *rays* generated by W.

What we have done explicitly in terms of canonical coordinates we can now do in a coordinate-free way as follows:

<u>Theorem 5.1</u>. Let $\phi: N \to M$ be a Hamilton-Jacobi submanifold of M, in the sense that it satisfies (5.2) and (5.3). Then, there is a unique vector field W_ϕ on N such that the following conditions are satisfied:

$$\phi_*(W_\phi) = V_H \quad \text{restricted to} \quad \phi(N) \qquad (5.8)$$

In particular, the image

$$t \to \phi(\sigma(t))$$

of an orbit curve $t \to \sigma(t)$ of W_ϕ is an orbit curve of V_H, i.e., is a solution of the Hamilton equations with Hamiltonian H.

<u>Proof</u>. One can regard the calculations in the local coordinates given above as a "proof", since local coordinates can always be chosen. However, the coordinate-free proof is even simpler, and goes as follows: The tangent vectors to a point x of M such that

$$v \rfloor \omega = (\text{scalar multiplier of } H)$$

form a one-dimensional linear subspace of M_x.

Now, let us see how the Liouville idea can be formulated in a coordinate-free way. Let us suppose that X is a product:

$$X = X_1 \times X_2 . \qquad (5.9)$$

Denote a point x of X by a pair

$$x = (x_1, x_2)$$

$$x_1 \in X_1, \quad x_2 \in X_2$$

Suppose that:

$$H(x_1, x_2) = \frac{H_1(x_1) + H_2(x_2)}{A_1(x_1) + A_2(x_2)} \tag{5.10}$$

where

$$x_1 \to H_1(x_1), A_1(x_1)$$

$$x_2 \to H_2(x_2), A_2(x_2)$$

are smooth functions on X_1 and X_2, respectively. An H of the form (5.10) will be said to be of *Liouville form* relative to the decomposition (5.9).

Let Y be a product

$$Y = Y_1 \times Y_2 \quad.$$

Let

$$\phi_1 : Y_1 \to M_1$$

$$y_1 \to \phi_1(y_1)$$

$$\phi_2 : Y_2 \to M_2$$

$$y_2 \to \phi_2(y_1)$$

be submanifold maps. Let

$$\phi \equiv \phi_1 \times \phi_2 : Y_1 \times Y_2 \to M_1 \times M_2 = M$$

$$(y_1, y_2) \to (\phi_1(y_1), \phi_2(y_2)) \quad. \tag{5.11}$$

Let us look for the condition that ϕ_1 and ϕ_2 must satisfy in order that the following condition be satisfied:

$$\phi^*(H) = E \tag{5.12}$$

where E is a real constant. Combine (5.10) with (5.11)

$$H_1(\phi_1(y_1)) + H_2(\phi_2(y_2)) = E(A_1(\phi_1(y_1)) + A_2(\phi_2(y_2)))$$

or

$$\phi_1^*(H_1 - EA_1)(y_1) = -\phi_2^*(H_2 - EA_2)(y_2) \qquad (5.13)$$

for all $y_1 \in Y_1$, $y_2 \in Y_2$.

Condition (5.13) is now, using the logic above, equivalent to the condition

$$\phi_1^*(H_1 - EA_1) = E_1$$
$$\phi_2^*(H_2 - EA_2) = -E_1 \qquad (5.14)$$

for some $E_1 \in R$.

The argument which led from (5.12) to (5.14) is reversible: If ϕ_1, ϕ_2 satisfy (5.14), and if ϕ is defined by (5.11), then it satisfies (5.12).

Now, suppose in addition that X_1 and X_2 are symplectic manifolds, with symplectic form ω_1 and ω_2 and that

$$\omega = \omega_1 + \omega_2 . \qquad (5.15)$$

Remark. Of course, (5.15) is shorthand for the formula

$$\omega = \pi_1^*(\omega_1) + \pi_2^*(\omega_2) \qquad (5.16)$$

where

$$\pi_1: M \equiv M_1 \times M_2 \to M_1$$

$$\pi_2: M \equiv M_1 \times M_2 \to M_2$$

are the Cartesian projections.

We can now formulate Liouville's condition in a coordinate free, and "global" form:

Theorem 5.2. Suppose X_1 and X_2 are symplectic manifolds with

$$H_1, A_1: X_1 \to R$$

$$H_2, A_2: X_2 \to R$$

smooth functions. Set:

$$X_1 = X_1 \times X_2 \qquad (5.17)$$

$$H = \frac{H_1 + H_2}{A_1 + A_2} \ . \tag{5.18}$$

The symplectic form on M is the sum (5.16) of the forms on X_1 and X_2.

For $E, E_1 \in R$, let

$$\phi_1^{E,E_1}: N_1 \to M_1$$

$$\phi_2^{E,E_1}: N_2 \to M_2$$

be submanifold maps such that:

$$(\phi_1^{E,E_1})^*(\omega_1) = 0 \tag{5.19}$$

$$(\phi_2^{E,E_1})^*(\omega_2) = 0 \tag{5.20}$$

$$(\phi_1^{E,E_1})^*(H_1 + EA_1) = E_1 \tag{5.21}$$

$$(\phi_2^{E,E_1})^*(H_2 + EA_2) = -E_1 \tag{5.22}$$

Set:

$$\phi^{E,E_1} = \phi_1^{E,E_1} \times \phi_2^{E,E_1} \tag{5.23}$$

Then,

$$\phi^{E,E_1}: N \equiv N_1 \times N_2 \to M$$

is a Hamilton-Jacobi map relative to the symplectic form ω and the Hamilton function H.

Further, let W_1^{E,E_1}, W_2^{E,E_1} be the vector fields on N_1 and N_2, respectively, such that

$$(\phi^{E,E_1})_*(W_1^{E,E_1}) \lrcorner \omega_1 = d(H_1 + EA_2) \tag{5.24}$$

$$(\phi^{E,E_1})_*(W_2^{E,E_1}) \lrcorner \omega_2 = d(H_2 + EA_2)$$

Set

$$W^{(E,E_1)} = W_1^{(E,E_1)} + W_2^{(E,E_1)} \tag{5.25}$$

a vector field in $N = N_1 \times N_2$. (Strictly, $W^{(E,E_1)}$ is the vector field on $N = N_1 \times N_2$ such that

$$\pi_*(W^{(E,E_1)}) = W_1^{(E,E_1)}$$

$$\pi_*(W^{(E,E_1)}) = W_2^{(E,E_1)})$$

If

$$t \to y(t)(y_1(t), y_2(t))$$

is an orbit curve of W^{E,E_1}, with $t \to y_1(t)$, an orbit curve of W^{E,E_1} and $t \to y_2(t)$ an orbit curve of W_2^{E,E_1}, then

$$t \to (y(t))$$

is a solution of Hamilton's equation with Hamiltonian H. In this way, the Hamilton equations are *reduced* to Hamilton's equations on spaces with a lower number of dimensions.

6. PLANAR LIOUVILLE SYSTEMS IN POLAR COORDINATES

A simple situation where one encounters "integrability" in both the Liouville and Lie forms is the problem of motion on the plane under central forces. This is the simplest general problem which can be treated with both Lie and Jacobi-Liouville theory. In this case (and other generalizations) one may hope to apply algebraic geometry (i.e., the theory of "Abelian integrals") and Lie theory to study more *global* properties of the Hamilton-Jacobi-Liouville method. In this section I will take preliminary steps toward working out this direction. Note also that much of the motivation for Darboux' *Theorie des Surfaces* is precisely that of determining "surfaces" (and associated mechanical systems) whose geodesic equations are of Liouville type in *specified coordinate systems*. This work has, in my opinion, a great resonance with the work in the last ten years in the applied mathematics and mathematical physics literature on "integrable" physical systems.

Note: In this section, subscripts do not denote partial derivations.

Suppose that

$$Q = R^2 - (0)$$

$$M = T^d(Q) \text{ the } cotangent\ bundle \text{ of } Q$$

with the usual symplectic structure. Let

(x,y)

be Cartesian coordinates for Q, (p_x, p_y) the corresponding momentum coordinates

$$w = d(p_x dx + p_y dy)$$

is the symplectic form.

Let (r, θ) be *polar coordinates*:

$$\begin{aligned} x &= r \cos \theta \\ y &= r \sin \theta \end{aligned} \quad . \tag{6.1}$$

Then, the following relation essentially defines the corresponding momenta (p_r, p_θ):

$$p_x dx + p_y dy = p_r dr + p_\theta d\theta \tag{6.2}$$

Use (6.1):

$$\begin{aligned} dx &= dr \cos \theta - r \sin \theta \, d\theta \\ dy &= dr \sin \theta + r \cos \theta \, d\theta \end{aligned} \quad . \tag{6.3}$$

Insert (6.3) into (6.2):

$$p_x(dr \cos \theta - r \sin \theta \, d\theta) + p_y(dr \sin \theta + r \cos \theta \, d\theta)$$

$$= p_r dr + p_\theta d\theta$$

or

$$\begin{aligned} p_r &= \cos \theta \, p_x + \sin \theta \, p_y \\ p_\theta &= r \cos \theta \, p_y - r \sin \theta \, p_x \end{aligned} \quad . \tag{6.4}$$

Then,

$$p_r^2 = \cos^2 \theta \, p_x^2 + \sin^2 \theta \, p_y^2 + 2 \cos \theta \sin \theta \, p_x p_y$$

$$p_\theta^2 = r^2 \cos^2 \theta \, p_y^2 + r^2 \sin^2 \theta \, p_x^2 - 2r^2 \cos \theta \sin \theta \, p_x p_y$$

$$r^2 p_r^2 + p_\theta^2 = 2r^2(p_x^2 + p_y^2) \tag{6.5}$$

Hamilton-Jacobi

Consider Hamiltonians of *Newtonian type*, which one can take of the following normalized form:

$$H(x,y,p_x,p_y) = \frac{1}{2} g(x,y)^{-1}(p_x^2 + p_y^2) + v(x,y) \tag{6.6}$$

Thus, in polar coordinates

$$H = \frac{1}{2} g(r,\theta)^{-1} \frac{(r^2 p_r^2 + p_\theta^2)}{r^2} + v(r,\theta) \tag{6.7}$$

Let us look for the conditions that this be in Liouville form, i.e., that there be functions H_r, H_θ, A_r, A_θ such that

$$H = \frac{H_r(r,p_r) + H_\theta(\theta,p_\theta)}{A_r(r,p_r) + A_\theta(\theta,p_\theta)} \tag{6.8}$$

In order that (6.7) be of the Liouville form (6.8), we must have:

$$g(r,\theta) \equiv \text{constant}$$

(which is taken to be 1)

$$v(r,\theta) \equiv v(r) ,$$

hence:

$$\begin{aligned} H_r &= \frac{1}{2} r^2 p_r^2 + r^2 v(r) \\ H &= \frac{1}{2} p_\theta^2 \\ A_r &= r^2 \\ B_r &= 0 \ . \end{aligned} \tag{6.9}$$

Here is the physical interpretation of this:

<u>Theorem 6.1</u>. The only Hamiltonians of this form which are of Liouville type in *this specified local coordinate system* represent, physically, central forces. From the point of view of Lie theory, they are the ones which have as symmetries a two-parameter abelian Lie algebra, namely, that generated by the vector fields

$$\frac{\partial}{\partial t} ,$$

$$H_{p_r}\frac{\partial}{\partial r} + H_{p_\theta}\frac{\partial}{\partial \theta} - H_r\frac{\partial}{\partial p_r} - H_\theta\frac{\partial}{\partial p_\theta}$$

We can now consider the *Hamilton-Jacobi-Liouville* partial differential equations:

$$H_r\left(r, \frac{dS_1}{dr}\right) - EA_r\left(r, \frac{dS_1}{dr}\right) = E_1$$

$$H_\theta\left(r, \frac{dS_2}{d\theta}\right) - EA_\theta\left(\theta, \frac{dS_2}{d\theta}\right) = E_1$$

(6.10)

As described in Section 4, find the general solution of these equations--as (E, E_1) vary--will determine the *general solution* of the Hamilton equations with the Hamiltonian H. Now, the equations (6.10) are of seperable type. If they can be solved in the following form

$$\frac{dS_1}{dr} = F_1(rp, E, E_1)$$

$$\frac{dS_2}{d\theta} = F_2(\theta p, E, E_1)$$

(6.11)

the solution can be written "by quadratures".

$$S_1 = \int F_1(r, E, E_1)\, dr$$

$$S_2 = \int F_2(\theta, E, E_1)\, d\theta$$

(6.12)

In addition, the right hand sides might have algebraic singularities, i.e., the integrals (6.12) might be what the classical algebraic geometers called *Abelian integrals*. The dependence of the integrals and the "periods" on the parameters E, E_1 involves *moduli* and *Picard-Lefschetz theory*.

Of course, in this central force case, the situation is relatively simple. Combining (6.10) and (6.9) we have:

$$\frac{1}{2}r^2\left(\frac{dS_r}{dr}\right)^2 + r^2 v(r) - Er^2 = E_1$$

(6.13)

$$\frac{1}{2}\left(\frac{dS_\theta}{d\theta}\right)^2 = -E_1.$$

(6.14)

For example, if $r \to v(t)$ is a rational function in r, solving (6.13) involves integrating a holomorphic differential form (with singularities) in a

Hamilton-Jacobi

hyperelliptic Riemann surface. The "elliptic" case, i.e., $r \to v(r)$ of degree ≤ 5, is especially important. In fact, note that Whittaker has already considered this type quite extensively in *Analytical Dynamics*.

Thus, we see that the theory of "integral" mechanical systems is a Happy Hunting Ground for the application of the theory of integration-of-holomorphism-with-singularities-differential-forms on algebraic varieties.

Of course, this Liouville point of view is not the only one: There is also the happy "coincidence" that the central force Hamiltonians can be attacked using both the Lie and Jacobi method.

Remark. Note the following result proved by Pars [3, Chapter 17] that the only (Newtonian) mechanical systems with two degrees of freedom that are *separable* (in the sense that the Hamilton-Jacobi equation has a complete solution which is a sum) are the Liouville systems.

Another type of interesting and seminal (from the point of view of *geometry* and *integrability* theory) mechanical system is the *rotating rigid body*. From this Hamilton-Jacobi-Liouville point of view, a generalization of the set-up described in previous sections, is the "nonholomonic" or "moving frame". I will now briefly consider this.

7. HAMILTONIAN SYSTEMS IN TERMS OF "NONHOLOMONIC" OR CARTAN MOVING FRAMES

The traditional geometric framework for treatment of Hamiltonian mechanics is in terms of a space of dimension $2n$, with distinguished coordinate systems (q^i, p_i), $1 \leq i, j \leq n$, such that the Hamiltonian structure is tied invariantly to the closed two-form:

$$dp_i \wedge dq^i .$$

This set-up is adequate for most purposes of *particle* mechanics, but becomes very awkward for geometrically more complicated situations that are often encountered in engineering, e.g., the *rotating rigid body*. I will now describe a symplectic formalism for dealing with such problems. My aim is to prepare the way for discussing *separability* for the Hamilton-Jacobi sense in this framework.

Let X continue as a symplectic manifold of dimension $2n$,

$$1 \leq i, j \leq n .$$

We shall consider representations of the symplectic form ω of the following type:

$$\omega = d(p_i \theta^i) , \qquad (7.1)$$

where

$$p_i \in \mathcal{F}(X) \equiv \mathcal{D}^0(X)$$

$$\theta^i \in \mathcal{D}^1(X) \equiv \text{one-differential forms on } M.$$

ω is the *symplectic form*, i.e.,

$$d\omega = 0 \qquad (7.2)$$

and

$$\underbrace{\omega \wedge \ldots \wedge \omega}_{n \text{ times}} \neq 0 \qquad (7.3)$$

$$d\theta^i = a^i_{jk} \theta^j \wedge \theta^k \qquad (7.4)$$

$$a^i_{jk} \in \mathcal{F}(M).$$

<u>Definition.</u> (7.1), with conditions (7.2)-(7.4), will be said to be *non-holonomic* or (Cartan) *moving frame representation* of the symplectic structure.

<u>Remark.</u> This can also be considered naturally from the theory of what Lie called "function groups". However, I usually prefer to follow Cartan's approach here. A common situation is that where the (a^i_{jk}) in (7.4) are *constants*. They are then *structure constants* of a Lie algebra, and (7.4) (considered as a set of differential equations for the θ^i) is called the *Cartan-Maurer equation*.

Given a function H, the *Hamiltonian*, form the vector field V_H such that:

$$V_H \lrcorner \omega = -dH. \qquad (7.5)$$

The *orbit curves*

$$t \to \exp(tV_H)(x_0)$$

of V_H (or the *orbits of the one-parameter pseudogroup generated by* V_H) are then the *solutions of the Hamilton equations with Hamiltonian* H. If Y is an n-dimensional manifold, if

$$\phi: Y \to X$$

Hamilton-Jacobi

is a submanifold map such that

$$\phi^*(\omega) = 0 \qquad (7.6)$$

$$\phi^*(H) = E, \text{ a constant,} \qquad (7.7)$$

then there is a unique vector field W on N such that:

$$\phi_*(W) = V_H .$$

Remark. Notice that I am changing notation slightly from that used previously, e.g., W now denotes a vector field rather than a solution of the Hamilton-Jacobi equation.

The curve

$$t \to (\exp(tW)(y))$$

$$y \in N ,$$

are *solutions* of the *Hamilton equations*. (7.6)-(7.7) is the Hamilton-Jacobi equation.

Suppose we look for ϕ satisfying the following *transversality* condition:

$$\phi^*(\theta^1 \wedge \ldots \wedge \theta^n) \neq 0 , \qquad (7.8)$$

i.e.,

The $\phi^*(\theta^i)$ are a basis for one-forms on N (7.9)

It follows from (7.6) that:

$$d\phi^*(p_i \theta^i) = 0 . \qquad (7.10)$$

Let us suppose there is a smooth function

$$y \to S(y)$$

on N such that:

$$dS = \phi^*(p_i \theta^i) . \qquad (7.11)$$

Theorem 7.1. If (7.8) is satisfied, then the map ϕ is (at least locally) uniquely determined by S (i.e., S up to an additive constant).

Proof. This follows from the fact that $\phi(N)$ is a *maximal integral* submanifold of ω.

8. LIOUVILLE SYSTEMS IN GENERAL GEOMETRIC TERMS

Just as before (which corresponds to the "holomonic" case $d\theta^i = 0$, i.e., $\theta^i = dq^i$), we can introduce the Liouville ansatz

If H has the following form

$$H = \frac{H_1 + \cdots + H_m}{A_1 + \cdots + A_m} \qquad (8.1)$$

find n-dimensional integral submanifolds

$$\phi^{(E,E_1,\ldots,E_m)} : N \to M$$

on which the following relations are satisfied:

$$\begin{aligned}
\phi^{(E,E_1,\ldots,E_m)*} (H_1 - EA_1) &= E_1 \\
&\vdots \\
\phi^{(E,E_1,\ldots,E_m)*} (H_m - EA_m) &= E_m
\end{aligned} \qquad (8.2)$$

Notice that (8.1) and (8.2) constitute an *overdetermined* system of first order *partial differential equations* for ϕ (or an exterior differential system parameterized by

$$(E, E_1, \ldots, E_m) \in R^{m+1}$$

with

$$E_1 + \cdots + E_m = 0 \ .$$

The compatibility conditions for these equations (i.e., that they be "in involution", in the sense of Lie and Cartan) are implied by the Liouville condition that X be (locally) describable as:

$$M_1 + \cdots + M_m \ ,$$

$$\omega = \omega_1 + \cdots + \omega_n \qquad (8.3)$$

$$\omega_a \in \mathcal{D}^1(M_a)$$

$$a = 1,\ldots,m$$

$$H_a, A_a \in \mathcal{F}(M_a)$$

Hamilton-Jacobi

In particular, notice that

$$0 = \{H_a, H_b\}$$
$$= \{H_a, A_b\}$$
$$= \{A_a, A_b\}$$

for $a \neq b$.

where $\{\,,\,\}$ is the Poisson bracket operation on $\mathscr{F}(X)$ defined by the symplectic form ω. Notice then that (8.4) can be regarded as the general *geometric* version of the Liouville conditions. This is a beautiful example of an exterior differential system *depending on parameters*!

9. MECHANICAL SYSTEMS WHOSE CONFIGURATION SPACE IS SO(3,R)

Just as the "simplest" systems in *particle* mechanics are those whose configuration space is R or R^2, so the simplest in *rigid body* mechanics are those whose configuration space is SO(3,R). Thus, many of the examples in the classical treatises (e.g., Whittaker [1], MacMillan [8], and Pars [3]) can be treated rationally and systematically from this point of view. In this section, I will present some material along these lines, from the Hamiltonian point of view.

Let X be a six dimensional symplectic manifold, whose symplectic form ω admits a representation of the following form:

$$\Omega = d(p_i \omega_i) \tag{9.1}$$

$$1 \leq i, j \leq 3$$

$$d\omega_1 = \omega_2 \wedge \omega_3$$
$$d\omega_2 = \omega_3 \wedge \omega_1 \tag{9.2}$$
$$d\omega_3 = \omega_1 \wedge \omega_2$$

(dp_i, ω_i) then form a basis for one-forms on X. (Of course, a "nonholomonic" one, since $d\omega_i \neq 0$.) One can obtain *local* coordinate systems by choosing

Euler angle realization of the $(\omega_1, \omega_2, \omega_3)$. In *Differential Geometry and the Calculus of Variations* I described how they may be considered from the viewpoint of general Lie group theory. (They are the natural coordinate systems for the symmetric space $SO(3,R) \times SO(3,R)/SO(3,R)$. Here I will simply take over the following formulas from the standard treatises:

$$-\omega_1 = \sin\theta \sin\phi \, d\psi + \cos\phi \, d\theta$$

$$-\omega_2 = \sin\theta \cos\phi \, d\psi - \sin\phi \, d\theta \qquad (9.3)$$

$$-\omega_3 = d\phi + \cos\theta \, d\psi \quad .$$

Theorem 9.1. If (θ, ϕ, ψ) are C^∞ functions in an open subset U of X, then $(\theta, \phi, \psi, p_1, p_2, p_3)$ are functionally independent, hence form a coordinate system after U is possibly made smaller.

Proof. This follows from general principles and the hypothesis that the two-form Ω defines a *symplectic* structure, i.e., has maximal rank.

Theorem 9.2. If (θ, ϕ, ψ) are a set of Euler angle functions for the forms (ω_i), then there exist (locally) functions p_θ, p_ϕ, p_ψ such that:

$$\Omega = dp_\theta \wedge d\theta + dp_\phi \wedge d\phi + dp_\psi \wedge d\psi \quad . \qquad (9.4)$$

The $(p_\theta, p_\phi, p_\psi)$ are linear combinations of the p_i, with coefficients which are functions of the (θ, ϕ, ψ) alone.

Proof. Let us impose the further condition that

$$p_1 \omega_1 + p_2 \omega_2 + p_3 \omega_3 = p_\theta d\theta + p_\phi d\phi + p_\psi d\psi \qquad (9.5)$$

Insert (9.3):

$-p_1 \sin\theta \sin\phi \, d\psi - p_1 \cos\phi \, d\theta$

$-p_2 \sin\theta \cos\phi \, d\psi + p_2 \sin\phi \, d\theta$

$-p_3 \, d\phi - p_3 \cos\theta \, d\psi$

$$= p_\theta d\theta + p_\phi d\phi + p_\psi d\psi \qquad (9.6)$$

Comparing coefficients of $d\theta, d\phi, d\psi$ on both sides of (9.6) give the required relations:

Hamilton-Jacobi

$$p_\theta = p_2 \sin \phi - p_1 \cos \phi$$

$$p_\phi = -p_3 \qquad (9.7)$$

$$p_\psi = -p_1 \sin \theta \sin \phi - p_2 \sin \theta \cos \phi$$

Suppose now that H is a function on X of the following form:

$$H = \frac{1}{2} I_{ij} p_i p_j + V, \qquad (9.8)$$

where (I_{ij}) is a 3×3 positive-definite, real and symmetric matrix (the *moment of inertia* matrix), and V is a real-valued C^∞ function. Consider H as the *Hamiltonian* of the mechanical system. There is then a vector field V_H on X such that:

$$V_H \lrcorner \Omega = -dH.$$

The orbits of the pseudogroup

$$t \to \exp(tV_H)$$

are then the basic geometric objects to be studied.

The explicit differential equations for the orbits can be found in a straightforward way in terms of the coordinates

$$(p_\theta, p_\phi, p_\psi, \theta, \phi, \psi):$$

$$\Omega = dp_\theta \wedge d\theta + dp_\phi \wedge d\phi + dp_\psi \wedge d\psi,$$

$$V_H \lrcorner \Omega = V_H(p_\theta)d\theta - V_H(\theta)dp_\theta \; V_H(p_\phi)d\phi - V_H(\phi)dp_\phi + V_H(p_\psi)d\psi - V_H(\phi)dp_\psi$$

$$dH = H_{p_\theta} dp_\theta + H_{p_\phi} dp_\phi + H_{p_\psi} dp_\psi + H_\theta d\theta + H_\phi d\phi + H_\psi d\psi$$

whence:

$$V_H(\theta) = H_{p_\theta}$$

$$V_H(p_\theta) = -H_\theta$$

$$V_H(\phi) = H_{p_\phi} \qquad (9.9)$$

$$V_H(p_\phi) = -H_\phi$$

$$V_H(\psi) = H_{p_\psi}$$

$$V_H(p_\psi) = -H_\psi$$

Thus, the orbit curves of V_H

$$t \to (\theta(t), \phi(t), \psi(t), p_\theta(t), p_\phi(t), p_\psi(t))$$

satisfy the *Hamilton equations:*

$$\frac{d\theta}{dt} = H_{p_\theta} \quad, \quad \frac{d\phi}{dt} = H_{p_\phi} \quad, \quad \frac{d\psi}{dt} = H_{p_\psi} \tag{9.10}$$

$$\frac{dp_\theta}{dt} = -H_\theta \quad, \quad \frac{dp_\phi}{dt} = -H_\phi \quad, \quad \frac{dp_\psi}{dt} = -H_\psi$$

This simplicity is bought at the price of making the formula for H complicated. Let us suppose, for example, that the moment of inertia matrix (I_{ij}) is diagonal (it can always be diagonalized of course by appropriate choice of the (ω_i))

$$H = \frac{1}{2}(I_1 p_1^2 + I_2 p_2^2 + I_3 p_3^2) + V \tag{9.11}$$

The formula for H in terms of the Euler angles and its conjugate momenta has been conveniently calculated by MacMillan [8, p. 379]. In the case

$$I_1 = I_2 \tag{9.12}$$

which we shall also assume:

$$H = \left[\frac{1}{2}\ I_1 p_\theta^2 + I_3 p_\phi^2 + I_1 \frac{(p_\psi - p_\phi \cos\theta)^2}{\sin^2\theta}\right] + V(\theta, \phi, \psi) \tag{9.13}$$

for $S(\theta, \phi, \psi)$ of the following form:

$$S(\theta, \phi, \psi) = S_1(\theta) + S_2(\phi) + S_3(\psi) \tag{9.14}$$

such that:

$$H\left(\theta, \phi, \psi, \frac{dS_1}{d\theta}, \frac{dS_2}{d\phi}, \frac{dS_3}{d\psi}\right) = E \ . \tag{9.15}$$

Hamilton-Jacobi

Thus, if S is a solution of (9.15), we can obtain a solution of (9.10) by solving the following ordinary differential equations:

$$\frac{d\theta}{dt} = H_{p_\theta}\left(\theta, \phi, \psi, \frac{\partial S}{\partial \theta}, \frac{\partial S}{\partial \phi}, \frac{\partial S}{\partial \psi}\right)$$

$$\frac{d\phi}{dt} = H_{p_\phi}\left(\theta, \phi, \psi, \frac{\partial S}{\partial \theta}, \frac{\partial S}{\partial \phi}, \frac{\partial S}{\partial \psi}\right) \quad (9.16)$$

$$\frac{d\psi}{dt} = H_{p_\psi}\left(\theta, \phi, \psi, \frac{\partial S}{\partial \theta}, \frac{\partial S}{\partial \phi}, \frac{\partial S}{\partial \psi}\right)$$

From (9.13) we can calculate the right hand side of (9.16):

$$H_{p_\theta} = I_1 p_\theta$$

$$H_{p_\phi} = I_1 (p_\phi + (p_\psi - p_\phi \cos\theta)) \left(-\frac{\cos\theta}{\sin^2\theta}\right) \quad (9.17)$$

$$H_{p_\psi} = I_3 p_\psi$$

Equations (9.16) now take the following form:

$$\frac{d\theta}{dt} = I_1 \frac{\partial S}{\partial \theta}$$

$$\frac{d\phi}{dt} = I_1 \left[\frac{\partial S}{\partial \phi} - \left(\frac{\partial S}{\partial \psi} - \frac{\partial S}{\partial \phi}\cos\theta\right)\frac{\cos\theta}{\sin^2\theta}\right] \quad (9.18)$$

$$\frac{d\psi}{dt} = I_3 \frac{\partial S}{\partial \psi}$$

Thus, if S is of the following *separated* form:

$$S(\theta, \phi, \psi) = S_1(\theta) + S_2(\phi) + S_3(\psi) \quad (9.19)$$

the equations (9.18) (which, recall, are projections of solutions of the Hamilton equations) take the following form:

$$\frac{d\theta}{dt} = I_1 \frac{dS_1}{d\theta} \quad (9.20)$$

$$\frac{d\phi}{dt} = I_1 \left(\frac{\frac{dS_2}{d\phi}}{\sin^2\theta} - \frac{dS_3}{d\psi}\frac{\cos\theta}{\sin^2\theta}\right) \quad (9.21)$$

$$\frac{d\psi}{dt} = I_3 \frac{dS_3}{d\psi} \tag{9.22}$$

(9.20) and (9.21) are in optimally separated form, and can be solved in terms of Abelian integrals:

$$\int \frac{d\theta}{\left(\frac{dS_1}{d\theta}\right)} = I_1 t \tag{9.23}$$

$$\int \frac{d\psi}{\left(\frac{dS_3}{d\psi}\right)} = I_3 t \tag{9.24}$$

After these Abelian integrals are inverted, (9.21) remains to be solved.

Of course, this possibility of exhibiting solutions of the Hamilton equations is conditional on a separated solution of the Hamilton-Jacobi equation existing. Let us examine this point. We must have:

$$\frac{1}{2}\left[I_1\left(\frac{dS_1}{d\theta}\right)^2 + I_3\frac{dS_2}{d\phi}^2 + I_1\frac{\left(\frac{dS_3}{d\psi} - \frac{dS_2}{d\phi}\cos\theta\right)^2}{\sin^2\theta}\right]$$

$$+ V(\theta,\phi,\psi) = E \tag{9.25}$$

We shall only consider the simplest situation, namely V a function of θ alone.

Theorem 9.2. Suppose that the potential V is a function of θ alone. Then, there is a separated solution of the Hamilton-Jacobi equation of the following form:

$$S(\theta,\phi,\psi) = a\phi + b\psi + S_1(\theta) \tag{9.26}$$

with a,b constant.

Proof. Substitute (9.26) into (9.25). Then,

$$\frac{1}{2}\left[I_1\left(\frac{dS_1}{d\theta}\right)^2 + I_3 a^2 + I_1\frac{(b - a\cos\theta)^2}{\sin^2\theta}\right] + V(\theta) = E$$

or

$$\frac{dS_1}{d\theta} = \left[2(E - V(\theta)) - I_1\left(\frac{b - a\cos\theta}{\sin^2\theta}\right)^2 - I_3 a^2\right]^{1/2} \tag{9.27}$$

Hamilton-Jacobi

$$\int \frac{d\theta}{\left[2(E-V(\theta)) - I_1\left(\frac{b - a\cos\theta}{\sin^2\theta}\right)^2 - I_3 a^2\right]^{1/2}} = I_1 t \qquad (9.28)$$

Suppose V is of the following form:

$$V(\theta) = \frac{p(\cos\theta)}{q(\cos\theta)} \qquad (9.29)$$

where $p(\ )$, $q(\ )$ are *polynomials*. Substituting

$$z = \cos\theta \qquad (9.30)$$

connects (9.28) into an Abelian integral over a *hyperelliptic curve*.

10. RIGID BODY MECHANICS AND THE NONHOLOMONIC MOVING FRAMES

In the end of the last section, we have reviewed some standard methods for the description, in the classical style, of some rigid body problems which are integrable in the Hamilton-Jacobi sense when viewed in terms of special coordinate systems for the configuration space $SO(3,R)$. Now, we turn to the description in terms of the moving frames

$$p_i, \quad \omega_i$$

for X, with:

$$\Omega = d(p_i \omega_i) \quad ,$$

$$d\omega_i = \varepsilon_{ijk} \omega_j \wedge \omega_k \quad .$$

These are more geometrically interesting, both for geometric and physical reasons. (For example, it leads to more elaborate Lie groups and systems which are integrable for more "modern" reasons than separation of variables.) Suppose that H is of the following form:

$$H = \frac{1}{2} I_{ij} p_i p_j + V \qquad (10.1)$$

$$dV = V_i \omega_i \quad . \qquad (10.2)$$

Let $\{\ ,\ \}$ denote the Poisson bracket operation defined by the symplectic form Ω.

$$dH = -V_H \mathbin{\lrcorner} \Omega$$

$$= -V_H \mathbin{\lrcorner} (dp_i \wedge \omega_i + p_i d\omega_i)$$

$$= -V_H(p_i)\omega_i + \omega_i(V_H) dp_i - 2p_i \epsilon_{ijk} \omega_j(V_H)_k \quad (10.3)$$

$$= \text{, using (10.1) and (10.2),}$$

$$I_{ij} p_i dp_j + V_i \omega_i \quad . \quad (10.4)$$

Comparing (10.3) and (10.4), we have the following equations:

$$\omega_i(V_H) = I_{ij} p_j \quad (10.5)$$

$$V_H(p_k) + 2 p_i \epsilon_{ijk} I_{j\ell} p_\ell = -V_K \quad (10.6)$$

We have proved the following:

<u>Theorem 10.1</u>. If $t \to x(t)$ is a curve in X which is a solution of the Hamilton equations with Hamiltonian H given by (10.1), then it satisfies the following differential equations:

$$\omega_i \frac{dx}{dt} = I_{ij} p_j (x(t)) \quad (10.7)$$

$$\frac{d}{dt} p_k(t) + 2\epsilon_{ijk} I_{j\ell} p_i p_\ell = -V_K(x(t)) \quad (10.8)$$

Further, if (q_j) are any set of functions on X such that:

$$\omega_i = a_{ij}(q) dq_j \quad (10.9)$$

(e.g., the Euler angle

$$q_1 = \theta; \quad q_2 = \phi, \quad q_3 = \psi \,)$$

then the Hamilton equations take the following form:

$$a_{ij}(q(t)) \frac{d}{dt}(q_j(t)) = I_{ij} p_j(t) \quad (10.10)$$

$$\frac{d}{dt} p_k(t) + 2\epsilon_{ijk} I_{j\ell} p_i(t) p_\ell(t) = -V_K(q(t)) \quad (10.11)$$

Equations (10.10)-(10.11) thus constitute a set of six nonlinear ordinary differential equations for the solutions of Hamilton's equations.

In the force-free case, i.e., $V = 0$, there is a remarkable simplification due to the decoupling of the 6×6 system. Equation (10.11), with $V_1 = 0$, constitutes a reduced system of three equations for the three unknowns $p_j(t) \equiv p_i(x(t))$. (They are called the *Euler equations of force-free rigid body motion*.) Whittaker [1] shows how to solve them explicitly in terms of the Jacobian elliptic functions. He also shows how the $t \to \theta(t), \phi(t), \psi(t)$ can be obtained in terms of the theta functions.

References

1. E. Whittaker, *Analytical Dynamics*, Cambridge Univ. Press, Cambridge, 1959.

2. F. Gantmacher, *Lectures in Analytical Mechanics*, Mir, Moscow, 1970.

3. L.A. Pars, *A Treatise on Analytical Dynamics*, Oxbow Press, Woodbridge, Conn., 1979.

4. V. Arnold, *USP. Mat. Nauk*. 18 (1963); Sur la géométrie différentielle des groupes de Lie de dimension infinie et ses applications à l'hydrodynamique des fluides parfaits, *Ann. Inst. Grenoble* 16, 319-361 (1966).

5. R. Hermann, The geometric theory of deformation and linearization of Pfaffian systems and its application to system theory and mathematical physics, I, *J. Math. Phys.* 24, 2268-2276 (1983).

6. R. Abraham and J.E. Marsden, *Foundations of Mechanics*, Addison-Wesley, Reading, MA, 1979, 2nd ed.

7. R. Hermann, Geometric construction and properties of some families of solutions of nonlinear partial differential equations, I, *J. Math. Phys.* 24, 510-521 (1983).

8. W.D. Macmillan, *Dynamics of Rigid Bodies*, Dover Publications, New York, 1960.

Chapter 2

INTEGRABLE HAMILTONIAN SYSTEMS
OF THE STÄCHEL TYPE

1. INTRODUCTION

In 1891 [1], Stächel, generalizing Liouville's work [2], found the (local) necessary and sufficient conditions that a Hamiltonian of the following form

$$H(p,q) = A_1(q)p_1^2 + \cdots + A_n(q)p_n^2 + V(q) \qquad (1.1)$$

separate in the sense that the associated Hamilton-Jacobi equation admits a *complete solution* of the form:

$$(q,a) \to S(q;a) = S_1(q_1;a) + \cdots + S_n(q_n;a) \qquad (1.2)$$

Once the Hamilton-Jacobi equation can be solved in this additive form, the Hamilton equations associated with the Hamiltonian (1.1) can be "solved" by quadratures and by finding inverse functions by an algorithm due to Jacobi. This is usually presented as a calculational matter. I want to emphasize here its geometric interpretation.

Stächel's "algorithm" briefly took a key importance for physics in the time (1905-1925) of the "old" quantum theory. The underlying physical systems could be "quantized" by algorithms due to Bohr and Sommerfeld. After the work of de Broglie, Heisenberg and Schrödinger transformed the "old" to the "new" quantum mechanics, Stächel's work faded into obscurity. It is now being revived because of its relation to the general theory of *integrable* physical systems. It is also of great interest to a geometer to ask how the Riemannian metric

$$ds^2 = A_1(q)^{-1}dq_1^2 + \cdots + A_n(q)^{-1}dq_n \qquad (1.3)$$

may be characterized "intrinsically" or in terms of embedding in Euclidean spaces. A basic problem motivating Darboux's 19th century treatise/masterpiece is the condition that the induced metric on a *surface* in R^3 be of Liouville type. Since, for n = 2, Stächel's conditions are precisely Liouville's, we can see the historic significance of these questions for differential geometry, independently of their relations to physics and engineering. However, to this day there is very little known about the conditions that a given Riemannian metric admit a decomposition of Stächel type.

Refer to Refs. 3-6 for the classical treatments of Stächel's theorem. (I have found Pars' treatment the most comprehensive and comprehensible.) In this paper, I want to emphasize another point of view.

2. THE STÄCHEL STRATEGY

Consider the Hamiltonian (1.1). Following Pars [3], p. 323, let us (changing notation slightly) ask whether there is a family of one-variable Hamiltonians

$$H_1(u,v; E_1,\ldots,E_n)$$
$$\vdots \qquad\qquad\qquad\qquad (2.1)$$
$$H_n(u,v; E_1,\ldots,E_n)$$

$(u,v) \in R^2$, $E_1,\ldots,E_n$ real parameters with the following property:

Let $S_1(u),\ldots,W_n(u)$ be solutions of the following differential equations:

$$H_1\!\left(u, \frac{dS_1}{du}\right) = \text{constant}$$
$$\vdots \qquad\qquad\qquad\qquad (2.2)$$
$$H_n\!\left(u, \frac{dS_n}{du}\right) = \text{constant}$$

Then

$$S(q_1,\ldots,q_n) = S_1(q_1) + \cdots + S_n(q_n)$$

is a *complete* solution (as $E_1,\ldots,E_n$ vary) of the Hamilton-Jacobi equation associated with the n-dimensional Hamiltonian (1.1).

In principle this method might be meaningful for very general Hamiltonians. (Geometrically, a "Hamiltonian" is a real-valued function on a symplectic manifold; usually the cotangent bundle to a configuration space manifold.) However, this conditions have only been worked out for the case (1.1), with a special form postulated for the one dimensional Hamiltonian $H_1,\ldots,H_n$.

Let us look for the $H_1,\ldots,H_n$ of the following form:

$$H_i(u,v; E_1,\ldots,E_n) = v^2 - \alpha_i^j(u)E_j - B_i(u) \qquad (2.3)$$

$i,j = 1,\ldots,n$. Summation over j in (2.3).

Stächel Type 35

$(A_i^j(u))$ is a matrix of functions of one variable *which is to be determined*. $E_1,\ldots,E_n$ are parameters, which are to be associated with the parameters needed to define a *complete* solution of the n-dimensional Hamilton-Jacobi equation. From the geometric point of view, the key point to Stächel's Ansatz is that the dependence on the parameters $E_1,\ldots,E_n$ is *linear*. Again, this indicates a certain deformation-theoretic foundation to the whole business!

Let us now find the relation between (2.3) and (1.1) in order that the Stächel Ansatz works. Let

$$S_1(u),\ldots,S_n(u)$$

be functions such that:

$$\begin{aligned} H_1\left(u, \frac{dS_1}{du}\right) &= \text{constant} \\ &\vdots \\ H_n\left(u, \frac{dS_n}{du}\right) &= \text{constant} \end{aligned} \tag{2.4}$$

Set:

$$S(q_1,\ldots,q_n) = S_1(q_1) + \cdots + S_n(q_n) \tag{2.5}$$

$$\frac{dS}{dq_i} = \frac{dS_i}{du}(q_i) \tag{2.6}$$

(no summation on i in (2.6). Suppose that:

$$\left(\frac{dS_i}{du}\right)^2 = \alpha_i^j(u)E_j + \beta_i(u) \tag{2.7}$$

Let S be defined by (2.5). Then,

$$\left(\frac{\partial S}{\partial q_i}\right)^2 = \text{, using (2.6)}$$

$$\left[\left(\frac{dS_i}{du}\right)(q_i)\right]^2$$

$$= \text{, using (2.7),}$$

$$\alpha_i^j(q_i)E_j + \beta_i(u_i) \tag{2.8}$$

(Again, no summation on i in (2.8).) Then,

$$A^i \left(\frac{\partial S}{\partial q_i}\right)^2 = A^i \alpha_i^j E_j + A^i \beta_i \ . \tag{2.9}$$

(Summation is now implied on j and i in (2.9).)

We can now sum up:

<u>Theorem 2.1</u>. Let $S_1(u),\ldots,S_n(u)$ be solutions of the one-dimensional Hamilton-Jacobi equations (2.7), with $(E_1,\ldots,E_n)$ considered (implicitly) as parameters. Then, $S: R^n \to R$ defined as *sum* (2.5) of the one-dimensional Hamilton-Jacobi functions is a solution of the Hamilton-Jacobi equation associated with the Hamiltonian (1.1) if and only if

$$A^i(q_1,\ldots,q_n)\alpha_i^j E_j + A^i \beta_i + V = \text{constant}$$

where α_i^j, β_i are considered functions on R^n as the pull-back under the projection $R^n \to R$.

Thus, we have the problem of deciding what conditions on the function A_i, V and R^n are implied by the conditions (2.10), where $\alpha_i(\ )$, $\beta_i(\ )$ are functions of one-variable alone.

3. THE STÄCHEL CONDITIONS

Consider (2.10) as a set of relations, with *variable* $(E_1,\ldots,E_n)$ between given functions $A^i(\ ,\ldots,\ )$, $V(\ ,\ldots,\ )$ on R^n and functions $\alpha_i^j(\ )$, $\beta_i(\ )$ on R. Consider the $(E_1,\ldots,E_n)$ as *independent variables*. (This is required by the condition that we end up with S as a *complete* solution of the Hamilton-Jacobi equation.) Relation (2.10) then forces the following relation:

$$A^i(q_1,\ldots,q_n)\hat{\alpha}_i^j \equiv \text{constant}$$
$$A^i(q_1,\ldots,q_n)\hat{\beta}_i + V \equiv \text{constant} \tag{3.1}$$

where $\hat{\alpha}_i^j, \hat{\beta}_i$ are functions of $(q_1,\ldots,q_n)$ of the form

$$\hat{\alpha}_i^j(q_1,\ldots,q_n) = \alpha_i^j(q_i)$$
$$\hat{\beta}_i(q_1,\ldots,q_n) = \beta_i(q_i) \tag{3.2}$$

(No summation on i in (3.2).)

Stächel Type   37

These are the classical Stächel conditions for separability of the Hamiltonian of form (1.1).

4. THE LIOUVILLE CONDITIONS AND CONFORMAL GEOMETRY

Let us now specialize the Hamiltonian (1.1) to one of the following form:

$$H = A(q_1,\ldots,q_n)(p_1^2 + \cdots + p_n^2) + V(q_1,\ldots,q_n) \quad . \tag{4.1}$$

The associated Riemannian metric

$$ds^2 = A^{-1}(dq_1^2 + \cdots + dq_n^2)$$

is then *conformal* to the flat metric. Let us see what the Stächel conditions (3.1) imply in this case.

Substitute the relations

$$A^i = A \tag{4.2}$$

for $i = 1,\ldots,n$

into (3.1):

$$A \sum_{i=1}^{n} \hat{\alpha}_i^j \equiv \text{constant}$$
$$\sum_{i=1}^{n} \hat{\beta}_i + V \equiv \text{constant} \tag{4.3}$$

Now, using (3.2)

$$A(q_1,\ldots,q_n) = \text{constant} \times (\alpha_1^j(q_1) + \cdots + \alpha_n^j(q_n))^{-1} \tag{4.4}$$

$$V(q_1,\ldots,q_n) = \text{constant} - (\beta_1(q_1) + \cdots + \beta_n(q_n)) \tag{4.5}$$

5. THE ORBITS OF THE STACHEL SYSTEMS AND THEIR QUADRATIC CONSERVATION LAWS

Let us return to a general Hamiltonian of Stächel type. Here are the formulas in compact form:

$$H(q,n) \equiv A_i^i p_i^2 + V$$

$$= \frac{1}{2} A^{ij} p_i p_j + V \tag{5.1}$$

with

$$A^{ij} = \begin{cases} 0 & \text{if } i \neq j \\ A^j & \text{if } i = j \end{cases}$$

$$H_i(u,v; E_1,\ldots,E_n) = v^2 - \alpha_i^j(u) E_j + \beta_i(u) \tag{5.2}$$

$$A^i(q) \hat{\alpha}_i^j(q) \equiv C^j(E_1,\ldots,E_n) \tag{5.3}$$

$$A^i(q) \hat{\beta}_i + V \equiv C(E_1,\ldots,E_n) \tag{5.4}$$

The $C, C^1,\ldots,C^n$ in (5.3)-(5.4) are functions of $E_1,\ldots,E_n$ alone.)

$$\hat{\alpha}_i^j(q_1,\ldots,q_n) = \alpha_i^j(q_i) \tag{5.5}$$

$$\hat{\beta}_i^j(q_1,\ldots,q_n) = \beta_i^j(q_i) \ .$$

Let $S_i(u)$ be solutions of the Hamilton-Jacobi equations for the Hamiltonian $H_1,\ldots,H_n$. Then solve for the following differential equations:

$$\left(\frac{dS_i}{du}\right)^2 = \alpha_i^j(u) E_j + \beta_i(u) \ , \tag{5.6}$$

or

$$S_i(u) = \int (\alpha_i^j(u) E_j + \beta_i(u))^{1/2} du \tag{5.7}$$

Set:

$$\hat{S}_i(q) = S_i(q_i) \quad \text{(no summation)} \tag{5.8}$$

$$S = S_1 + \cdots + S_n \ . \tag{5.9}$$

Then,

$$\frac{\partial S}{\partial q_i} = \frac{dS_i}{du}(q_i)$$

$$= \ , \text{ using (5.7)}$$

$$(\hat{\alpha}_i^j E_j + \hat{\beta}_i)^{1/2}$$

$$\frac{\partial H}{\partial p_i} = A^{ij} p_j \qquad \text{(no summation)} \qquad (5.10)$$

Hence, the curves

$$t \to (p(t), q(t))$$

which satisfy the following *ray equations* are solutions of Hamilton's equations:

$$p_i(t) = (\hat{\alpha}_i^j(q(t)) E_j + \hat{\beta}_i(q(t)))^{1/2} \qquad (5.11)$$

$$\frac{dq^i}{dt} = -\frac{\partial H}{\partial p_i}\left(q, \frac{\partial S}{\partial q_i}\right)$$

$$= A^{ij} \frac{\partial S}{\partial q_j}$$

$$= A^{ij} (\hat{\alpha}_j^k E_k + \hat{\beta}_j)^{1/2}$$

$$= A^i (\hat{\alpha}_i^j E_j + \hat{\beta}_i)^{1/2}$$

$$= (A^i)^{1/2} (A^i (\hat{\alpha}_i^j E_j + \hat{\beta}_i))^{1/2}$$

$$= (A^i)^{1/2} (C^j(E_1,\ldots,E_n) E_j + C(E_1,\ldots,E_n))^{1/2}$$

We can now use the Stachel relations (5.3)-(5.4) to simplify the ray equations (5.10). The right hand side of (5.10) can be written as:

$$2((A^j)^2 (\hat{\alpha}_i^j E_j + \hat{\beta}_i))^{1/2} = 2(A^i)^{1/2} (C^j + C)^{1/2} \quad .$$

<u>Theorem 5.1.</u> The rays are solutions of the following differential equations:

$$\frac{dq^i}{dt} = 2(A^i(q(t)))^{1/2} [C^j(E_1,\ldots,E_n) + E_j C(E_1,\ldots,E_n)]^{1/2} \qquad (5.12)$$

Geometrically, this means that the orbits are, after a *constant* change in time scale (which depends on the Jacobian "orbital elements" $E_1,\ldots,E_n$) solutions of the differential equations:

$$\frac{dq^i}{dt} = (A_i(q(t)))^{1/2} \quad . \qquad (5.13)$$

We can give another geometric interpretation of the result of Theorem 5.1 which is very important from the physical point of view. We can show that the Stachel systems admit n *conservation laws which are quadratic in the momenta*.

Let $t \to (q(t), p(t))$ be a solution of Hamilton's equation. We know (Jacobi's theorem and (5.11)) that for some choice of the "orbital elements" $(E_1, \ldots, E_n)$

$$p_i(t)^2 = \hat{\alpha}_i^j(q(t)) E_j + \hat{\beta}_i(q(t)) \tag{5.14}$$

Let (γ_i^j) be the inverse to the matrix $(\hat{\alpha}_i^j(q))$. Then, (5.14) can be rewritten as

$$\gamma_i^j(q(t))(p_j(t)^2 - \hat{\beta}_i(q(t))) \equiv \text{constant} \; .$$

Set:

$$f_i(q,p) = \gamma_i^j(q)(p_j^2 - \hat{\beta}_i(q)) \tag{5.15}$$

f_i is now *independent of the* E_j, hence we have:

$$\frac{d}{dt}(f_i(q(t), p(t))) = 0 \tag{5.16}$$

along *every* solution of Hamilton's equation with Hamiltonian H. But, the left hand side is the Poisson bracket $\{f_i, H\}$ with the Hamiltonian H. Hence, we have proved:

<u>Theorem 5.2</u>. The functions f_i of (q,p) (which, in global terms is a function on either the tangent or cotangent bundle to the manifold Q of variables q) is *conserved* under the Hamiltonian flow, in the sense that

$$\{f_i, H\} = 0 \tag{5.17}$$

where $\{ \, , \, \}$ is the *Poisson bracket*. In particular, f_i is a quadratic *Killing tensor* for the Riemannian metric

$$ds^2 = A_i^{-1} dq_i^2 \; .$$

Of course, functions f_i satisfying (5.1) are known in another guise as the *infinitesimal generators* of one parameter groups of *symmetries* of the Hamilton equations. From this Lie theoretic point of view, it is important to have information about the Poisson brackets $\{f_i, f_j\}$. We will show in Section 6 that they are *zero*, which ties in with "integrability" in the sense

Stächel Type 41

it is known in the work of Liouville, Lie and Arnold. We will call these conserved functions the *quadratic Stachel infinitesimal symmetries*.

6. THE STACHEL INFINITESIMAL SYMMETRIES FROM THE LIE THEORETIC AND SYMPLECTIC POINT OF VIEW

Let us go to a more general point of view. Let X be a 2n-dimensional symplectic manifold whose coordinates are

$$(p_i, q^i)$$

and whose symplectic form is

$$\omega = dp_i \wedge dq^i \qquad (6.1)$$

$$H = A^i(q) p_i^2 + V_i(q) . \qquad (6.2)$$

Let Q be the manifold whose coordinates are (q^i). A solution of the Hamilton-Jacobi equation is a submanifold map

$$\phi: Q \to X$$

such that:

$$\phi^*(\omega) = 0 \qquad (6.3)$$

and

$$\phi^*(dH) = 0 , \qquad (6.4)$$

i.e., ϕ is a *maximal integral submanifold* of the exterior differential system on X generated by ω and dH. The solutions of the Hamilton equations are the *Cauchy characteristic curves* of this system. (Locally,

$$\phi(q) = \left(q, \frac{\partial S}{\partial q}\right)$$

where $q \to S(q)$ is a solution of the Hamilton-Jacobi equation in the usual sense

$$H\left(q, \frac{\partial S}{\partial q}\right) \quad \text{constant} \quad .)$$

Let M be a two-dimensional symplectic manifold, with canonical coordinates (u,v). Let $H_i(u,v; E_1,\ldots,E_n)$ be the one-dimensional Hamiltonian involved in the Stächel construction.

$$H_i = v^2 - \alpha_i^j(u)E_j - \beta_i(u) \tag{6.5}$$

Let

$$\pi_i : X \to M$$

be the submersion map defined as follows:

$$\pi_1(q,p) = (q^1, p_1)$$
$$\vdots$$
$$\pi_n(q,p) = (q^n, p_n) \ .$$

Then,

$$\pi_i^*(v) = p_i$$
$$\pi_i^*(u) = q^i \ . \tag{6.6}$$

Let us normalize the choice of constants in (6.5) so that the Hamilton-Jacobi equation is:

$$f_i - \gamma_i^j \pi_j^*(H) = \text{constant} \ .$$

Hence, the exterior differential system generated by ω and the $df_1, \ldots, df_n$ is "in involution" in the sense of Lie. Hence, the Poisson brackets now satisfy the following condition:

$$d(\{f_i, f_j\}) \text{ belongs to the exterior differential ideal generated by } \omega \text{ and the } df_1, \ldots, df_n.$$

As the E's vary, one can, in fact, show that the Poisson bracket $\{f_i, f_j\}$ *vanish*. The explicit calculation is left to the reader. This situation suggests some further general geometric remarks.

7. ON THE RELATION BETWEEN INTEGRABILITY IN THE SENSE OF LIOUVILLE/LIE/ARNOLD AND ALGEBRAIC GEOMETRY

Let X continue as a $2n$ dimensional symplectic manifold, with canonical local coordinates (p_i, q^i). Let H be a Hamiltonian function and let (f') be a set of n functions on M such that the following conditions are satisfied:

$$\{H, f^i\} = 0$$
$$= \{f^i, f_j\} \tag{7.1}$$

$$df^1, \ldots, df^n \text{ are everywhere independent of } X. \tag{7.2}$$

i.e., the map

$$x \to f_1(x), \ldots, f_n(x) \tag{7.3}$$

of $X \to R^n$ is a *submersion*.

Theorem 7.1. There is, locally, a function $F(\ ,\ldots,\ )$ on R^n such that:

$$H = F(f_1, \ldots, f_n). \tag{7.4}$$

Proof. Left to the reader. Follows from general facts about symplectic manifolds.

For each

$$E = (E_1, \ldots, E_n) \in R^n$$

set

$$X_E = \{x \in X: f_1(x) = E_1, \ldots, f_n(x) = E_n\} \tag{7.5}$$

Theorem 7.2. For each E in the image of the submersion map (7.4), X_E is an n-dimensional submanifold of X. It is a so-called "Lagrangian" submanifold, in the sense that it is a maximal integral submanifold of the exterior differential system generated by ω. Locally, there are functions $q_1, \ldots, q_n$ such that

$$\omega = dq_1 \wedge df_1 + \cdots + dq_n \wedge df_n \tag{7.6}$$

Proof. Again, left to the reader, since it is standard symplectic manifold "Problem of Pfaff" material, done very elegantly and simply already (and in a completely "modern" way!) by Cartan and Goursat in the 1920's!

For $f \in \mathscr{F}(X)$, let V_f be the vector field such that:

$$V_f \lrcorner\, \omega = -df. \tag{7.7}$$

Theorem 7.3. The map

$$f \to V_f$$

is a Lie algebra homomorphism $\mathscr{F}(X) \to \mathscr{V}(X)$ where $\mathscr{F}(X)$ are the C^∞ real valued functions on X, made into a Lie algebra under Poisson bracket, and $\mathscr{V}(X)$ is a Lie algebra under commutation/Jacobi/Lie bracket.

Proof. Again, standard.

Theorem 7.4. Let $V_{f_1}, \ldots, V_{f_n}$ be the vector fields on X associated with $f_1, \ldots, f_n$ satisfying (7.1). Then,

$$[V_{f_1}\ V_{f_2}] = 0$$

V_{f_i} are tangent to the submanifolds X_E, and define an absolute parallelism on X_E.

The pseudogroup (in the Ehresmann sense) generated by $V_{f_1}, \ldots, V_{f_n}$ is:

a) Abelian.

b) Maps each submanifold X_E into itself.

c) Acts locally transitively on each X_E.

d) Globalizable, as a Lie group action of X_E is *compact*. X_E, in this case, is a *torus*, which corresponds to the "action angle variable" construction of the "old" 1905-25 quantum theory.

Proof. Again, standard.

Now, as the parameters $E = (E_1, \ldots, E_n)$ vary, we obtain a family of submanifolds X_E. This is, of course, a typical "deformations" situation. What is also being deformed is the abelian pseudogroup generated by the vector fields

$$V_{f_1}, \ldots, V_{f_n} \ .$$

Now, of course, a basic question for algebraic geometers is whether these tori have a *group* structure.

Now we ask for a more explicit connection between algebraic geometry, particularly the theory of "abelian integrals", the mechanics situation. After all, the very historical ongoings of algebraic geometry, in the theory of "elliptic integrals" of Euler, Legendre, Abel and Jacobi, are in this question!

Stächel Type

Let us look at the Stachel situation for enlightenment here: Suppose the f_i are of the following form:

$$f_i(p,q) = \gamma_i^j(\mathscr{P}_j^2 - \hat{\beta}_j) \tag{7.8}$$

with

$$\hat{\beta}_j(q_1,\ldots,q_n) = \beta_i(q_i) \tag{7.9}$$

$$\gamma_i^j(q)\hat{\alpha}_j^k(q) = \delta_i^k \tag{7.10}$$

$$\hat{\alpha}_j^k(q) = \alpha_j^k(q_j) . \tag{7.11}$$

We can then find a complete solution to the Hamilton-Jacobi equation $S(E_1,\ldots,E_n)$ of the following form:

$$S(q;E) = S_1(q_1,E) + \cdots + S_n(q_n,E) \tag{7.12}$$

with

$$\left(\frac{dS_i}{du}\right)^2 = \alpha_i^j(u)E_j + \beta_i(u) . \tag{7.13}$$

We then have a map

$$\phi: Q \to X_E$$

which takes the following form:

$$q \to \left(q, \frac{\partial S_1}{\partial q_1},\ldots,\frac{\partial S_n}{\partial q_n}\right)$$

$$= (q, [\alpha_1^j(q_1)E_j + \beta_1(q_1)]^{1/2},\ldots,[\alpha_n^j(q_n)E_j + \beta_n(q_n)]^{1/2} \tag{7.14}$$

These formulas exhibit X_E as a *real algebraic* variety (assuming that the α_i^j, β_i are algebraic functions of q). Further, this n-dimensional variety is the *product* of the one-dimensional ones:

$$\begin{aligned}\mathscr{A}_1 &= \{(u,v): v^2 = \alpha_1^j(u)E_j + \beta_1(u)\} \\ \mathscr{A}_n &= \{(u,v): v^2 = \alpha_n^j(u)E_j + \beta_n(u)\}\end{aligned} \tag{7.15}$$

For example, if the $\alpha_i^j(u)$, $\beta_i(u)$ are *polynomials* in u, $\mathscr{A}_1,\ldots,\mathscr{A}_n$ are *hyperelliptic elliptic curves*, defining them to be the algebraic varieties of the form

$$\{(u,v): v^2 = \sqrt{p(u)}\}$$

when $u \to p(u)$ is any polynomial in u.

References

1. P. Stäckel, J. Reine, *Ang. Math.* <u>107</u>, 319 (1891); <u>128</u>, 222 (1905).

2. P. Liouville, *Journal de Math.* XIV, 257 (1849).

3. F. Whittaker, *Analytical Dynamics*, Cambridge Univ. Press, 1919.

4. L.A. Pars, *A Treatise on Analytical Dynamics*, Oxbow Press, Woodbridge, Conn., 1979.

5. F. Gantmacher, *Lectures in Analytical Mechanics*, Mir, Moscow, 1970.

6. Y. Hagihara, *Celestial Mechanics*, Vol. 1, MIT Press, 1970.

PART II

GEOMETRIC STRUCTURES IN INTEGRABILITY THEORY

Having reviewed, in Part I, some of the 19th century mechanics related material, it is natural to ask what it means in terms of contemporary mathematics and physics. Indeed, this is the part of classical integrability theory which interests me!

Part of the meaning is algebro-geometric: Integrability for physico-mechanical systems implies (as Arnold has proved) in Lie theoretic terms a *torus* or *Abelian variety* structure, hence the possibility of explicit soluttions via theta functions. However, another "meaning" -- which I would trace back at least to Darboux' *Theorie des Surfaces* -- is the existence of some sort of *geometric structure* with certain properties. Now, I do not yet see this as a systematic mathematical theory, but I believe it could become one. In these next chapters I present some new ideas in this direction.

Chapter 3

THE CALOGERO EQUATIONS FOR INTEGRABLE SYSTEMS

Abstract

A general form of the Calogero equations for a certain type of Lax representation is presented. The objects involved are associated with certain vector bundles, and the conditions are described in terms of invariantly defined differential operators on these vector bundles.

1. INTRODUCTION

Toda [1] introduced a class of mechanical systems which are, in the terminology of Whittaker [2], "soluble". (The term "completely integrable" is now more commonly used, but conflicts with its traditional use in differential geometry as a property of exterior differential equations.) This work started an extensive area of research. Flaschka [3] showed how Toda's system could be put into Lax form [4]. Calogero then provided [5] a Lax representation for a more extended class of one dimensional lattice-like systems, and introduced a more systematic way of searching for them. The aim of this paper is to analyze the differential-geometric reasoning that goes into Toda-Flaschka-Calogero calculations and continue the work of this series [6] on the geometric foundation of the *integrability* property of physical systems. In this paper, I will take a different approach from that in Ref. 6 toward the geometric interpretation of the "Lax representation" of a physical system: I will look for the Lax representation as a *solution of certain differential equations*.

Suppose

$$q = (q^i), \quad p = (p_j), \quad 1 \leq i, j \leq n$$

summation convention

are the *position* and *momentum* coordinates of a mechanical system. Let X be a region of R^{2n}, considered as a differentiable manifold with coordinates (q,p). Let

$$\{\,,\,\}: \mathscr{F}(X) \times \mathscr{F}(X) \to \mathscr{F}(X)$$

be the Lie algebra operation

$$(f_1, f_2) \to \frac{\partial f_1}{\partial p_i} \frac{\partial f_2}{\partial q^i} - \frac{\partial f_1}{\partial q^i} \frac{\partial f_2}{\partial p_i} \tag{1.1}$$

of *Poisson bracket*. ($\mathscr{F}(X)$ = the C^∞, real-valued functions on X.) Let $h \in \mathscr{F}(X)$ be the *Hamiltonian function* of a mechanical system. One associates with h the *Hamilton equations*

$$\frac{dq^i}{dt} = \frac{\partial h}{\partial p_i}$$

$$\frac{dp_i}{dt} = -\frac{\partial h}{\partial q^i} \quad . \tag{1.2}$$

In this paper, a *Lax representation* for the equations (1.2) will mean a pair

$$\underline{L} = (L_i^j) \,, \qquad \underline{M} = (M_i^j)$$

of $n \times n$ matrices whose entries are elements of $\mathscr{F}(X)$ such that the following equations are satisfied:

$$\{h, L_i^j\} = M_i^k L_k^j - L_i^k M_k^j \quad . \tag{1.3}$$

In his work [5], Calogero looked for h's, L's, and M's of specified form, and described the elements of this form in terms of certain ordinary differential equations. (Ultimately, these equations could be solved by means of elliptic functions.) In this paper I will analyze this reasoning in a wider geometric context.

What is especially noteworthy is that the Calogero condition can be interpreted as curvature conditions for a connection in a fiber bundle sitting over the space of the q-variables.

2. THE CALOGERO EQUATIONS

Let $h(p,q)$ be the Hamiltonian function of the mechanical system. We shall suppose it to be of the following form:

$$h = \frac{1}{2} g^{ij}(q) p_i p_j + V(q) \tag{2.1}$$

where (g^{ij}) is a symmetric non-degenerate matrix of C^∞ functions of q, and $V(q)$ is a C^∞ function of q. If $f(p,q)$ is any C^∞ function of p and q, let $\{h, f\}$ be the Poisson bracket given by formula (1.1). Let $1 \le a, b, \ldots \le m$ be a new set of indices, with the corresponding summation convention.

We shall look for a set of functions of (p,q) labelled as follows:

$$(L_a, M_a^b) \,, \tag{2.2}$$

Calogero Equations 51

which satisfy the following *generalized Lax equations*

$$\{h, L_a\} = M_a^b L_b ,\qquad (2.3)$$

and in addition, the following differential equations:

$$\{p_i, \{p_j, M_a^b\}\} = \{p_i, \{p_j, L_a\}\} = 0 \qquad (2.4)$$

(2.4) implies that the L_a and M_a^b are of the following form:

$$L_a = \alpha_a^i p_i + \beta_a \qquad (2.5)$$

$$M_a^b = \gamma_a^{bi} p_i + \sigma_a^b , \qquad (2.6)$$

where α_a^i, β_a, γ_a^{bi}, σ_a^b are functions of q alone. One can readily derive the following equations for the $\alpha, \beta, \gamma, \sigma$:

$$g^{ij}\frac{\partial \alpha_a}{\partial q^i} + g^{ik}\frac{\partial \alpha_a^j}{\partial q^i} - \frac{\partial g^{jk}}{\partial q^i}\alpha_a^i = \gamma_a^{bj}\alpha_b^k + \gamma_a^{bk}\alpha_b^j \qquad (2.7)$$

$$g^{ij}\frac{\partial \beta_a}{\partial q^i} = \gamma_a^{bj}\beta_b + \sigma_a^b \alpha_b^j \qquad (2.8)$$

$$-\frac{\partial V}{\partial q^i}\alpha_a^i = \sigma_a^b \beta_b . \qquad (2.9)$$

Regarding α_a^k, β_a as unknown functions in (2.7)-(2.8), there are certain compatibility conditions on the coefficients (γ, σ) required for the existence of a solution. We will call them the *Calogero equations*. Equations (2.9) are an additional set of conditions which much be satisfied.

3. THE CALOGERO EQUATIONS AS CURVATURE CONDITIONS

We will now integrate Equations (2.7)-(2.8) in terms of the theory of connections in vector bundles [7]. Let

$$(g_{ij}) \qquad (3.1)$$

be the inverse matrix to (g^{ij}). Multiply both sides of (2.7)-(2.8) by

$$g_{\ell j}$$

to obtain the following equations:

$$\frac{\partial \alpha_a^k}{\partial q^\ell} + \frac{\partial \alpha_a^j}{\partial q^k} - g_{\ell j}\frac{\partial q^{jk}}{\partial q^i}\alpha_a^i \;=\; g_{\ell j}\gamma_a^{bj}\alpha_b^k + g_{\ell j}\gamma_a^{bk}\alpha_b^j \qquad (3.2)$$

$$\frac{\partial \beta_a}{\partial q^\ell} \;=\; g_{\ell j}\gamma_a^{bj}\beta_b + g_{\ell j}\sigma_a^{b}\alpha_b^j \;. \qquad (3.3)$$

Let us now introduce a manifold Q whose coordinates are (q^i) (so that the space of variables (q,p) may be identified naturally with the cotangent bundle to q), and let E be the space of variables

$$(q^i, \beta_a) \;. \qquad (3.4)$$

Let $\pi : E \to Q$ be the map

$$\pi(q,\beta) \;=\; q \;. \qquad (3.5)$$

π defines E as a *vector bundle* over Q of real, m-dimensional vector spaces. The cross-section maps

$$\Gamma(E)$$

can then be identified as the space of maps

$$\beta : Q \to E$$

of the form:

$$\beta(q) \;=\; (q, \beta_a(q)) \qquad (3.6)$$

where $(\beta_a(q))$ are a collection of m real-valued functions.
Let

$$E \otimes T(Q)$$

be the tensor product of E with the *tangent vector bundle* to Q. Its cross sections

$$(E \otimes T(Q))$$

can be identified with the space of functions

$$(\alpha_a^k(q)) \;.$$

We can use the $(g_{\ell j})$ and (γ_a^{bj}) to define a linear connection ∇ in E. For each vector field V, ∇_V is to be a linear, first order differential operator: $\Gamma(E) \to \Gamma(E)$, which depends "tensorially" on V, i.e., the map

Calogero Equations 53

$V \to \nabla_V$ is $\mathscr{F}(Q)$-linear. Thus, $V \to \nabla_V$ is determined by its values on the coordinate vector fields: $\partial/\partial q^i$.

We can define a basis (e^a) of the $\mathscr{F}(Q)$-module $\Gamma(E)$ as follows:

$$e^a(q) = (q^1(q),\ldots,q^n(q), 0,\ldots,1,\ldots,0) \qquad (3.7)$$

where, on the right hand side of (3.7), the "1" is in the a-th place. Thus, if $\beta \in \Gamma(E)$ is given as follows:

$$\beta(q) = (q^1(q),\ldots,q^n(q), \beta_1(q),\ldots,\beta_m(q)) \qquad (3.8)$$

then:

$$\beta = \beta_a e^a . \qquad (3.9)$$

With this algebraic notation, we can now define the linear connection ∇ in the vector bundle E by the following formula:

$$\nabla_{\frac{\partial}{\partial q^i}} (\beta) = \left(\frac{\partial \beta_a}{\partial q^i} + g_{ij} \gamma_a^{bj} \beta_b\right) e^a . \qquad (3.10)$$

Consider the Riemannian metric

$$ds^2 = g_{ij} dx^i dx^j \qquad (3.11)$$

as fixed. ∇ then depends on the choice of the γ's.

Notice that the covariant derivative of γ itself is a cross-section of

$$E \otimes T^d(Q) , \qquad (3.12)$$

the tensor product of the tangent and cotangent bundle. Use the metric (3.11) to identify $T^d(Q)$ with $T(Q)$. The left hand side of (3.2) then is given invariantly by a first order linear differential operator

$$D: \Gamma(E \otimes T(Q)) \to \Gamma(E \otimes (T(Q) \circ T(Q)) \qquad (3.13)$$

($\circ$ denotes symmetric tensor product.) The right hand isde is a zero-th order linear differential operator

$$A: \Gamma(E \otimes T(Q)) \to \Gamma(E \otimes (T(Q) \circ T(Q))) \qquad (3.14)$$

(3.11) then takes the form:

$$D\alpha = A\alpha \qquad (3.15)$$

with

$$\alpha = \alpha_a^k e^a \otimes \frac{\partial}{\partial q^k} \quad . \tag{3.16}$$

Let σ be the zeroth-order linear differential operator

$$\sigma: \Gamma(E \otimes T(Q)) \to \Gamma(E \otimes T(Q))$$

given as follows:

$$\sigma\left(\alpha_a^i e^a \otimes \frac{\partial}{\partial q^i}\right) = \left(\sigma_{ab}^b \alpha_b^j \frac{\partial}{\partial q^j}\right) e^a \quad . \tag{3.17}$$

Then, Equation (3.3) takes the form:

$$\nabla \beta = \sigma(\alpha) \quad . \tag{3.18}$$

Let us sum up as follows:

<u>Theorem 3.1</u>. The first two of the Calogero equations (2.7)-(2.8) take the coordinate-free form:

$$\begin{aligned} D\alpha &= A\alpha \\ \nabla \beta &= \sigma(\alpha) \end{aligned} \quad , \tag{3.19}$$

where D, ∇ are first-order linear differential operators on vector bundles, σ and A are zero-th order operators. D is an operator which depends only on the metric (3.3), while all the rest depend on the choice of the functions $(\sigma_a^b, \gamma_a^{bi})$. Thus, *involutiveness* of these equations, in the classical sense [8-11], forces differential equations on the $(\sigma_a^b, \gamma_a^{bi})$ which includes the Calogero theory. The compatibility of these equations (2.9) then determines conditions on the potential function V.

<u>Remark</u>. One can alternately regard (2.7)-(2.8) as a system to be solved for the (α_a^i, β_a, V). The classical condition that this system be *in involution* [8-11] then determines the differential equations for the σ_a^b and γ_a^{bj}.

4. THE CALOGERO EQUATIONS IN A MORE GENERAL SETTING

We can now abstract out of the specific context of Calogero's papers some further general remarks. Let X be a C^∞, paracompact manifold, and let $\mathcal{V}(X)$ be the C^∞ vector fields on X. Suppose $V \in \mathcal{V}(X)$ is given. Its orbit curves $t \to x(t)$ satisfy ordinary differential equations

$$\frac{dx}{dt} = V(x(t)) \quad . \tag{4.1}$$

Calogero Equations

Let $\mathcal{F}(X)$ be the C^∞, real-valued functions on X, a commutative associative algebra. (Thus, $\mathcal{V}(X)$ is the Lie algebra of derivations of $\mathcal{F}(X)$.) Choose the following range of indices and the summation convention:

$$1 \leq a, b \leq m \; .$$

Definition. A set labelled (L_a, M_a^b) of elements of $\mathcal{F}(X)$ is a solution of the *Lax-Calogero equations* if (relative to the vector field V) the following conditions are satisfied:

$$V(L_a) = M_a^b L_b \; . \tag{4.2}$$

The following result follows from the differential geometric meaning of things and shows how solutions (L,M) of (4.2) can be used to generate constants of motion of V.

Theorem 4.1. Let $\mathcal{G}$ be the Lie algebra of $m \times m$ matrices generated by the matrices of the form:

$$\{M_a^b(x) : x \in X\} \; .$$

Let G be the subgroup of GL(m,R), the Lie group of non-singular $m \times m$ real matrices, generated by the elements of $\mathcal{G}$. Let $F \in \mathcal{F}(R^m)$ be a real-valued function on R^m, which is invariant under the action of G on R^m. Define a function $f \in \mathcal{F}(X)$ as follows:

$$f(x) = F(L_1(x), \ldots, L_n(x)) \tag{4.3}$$

for $x \in X$.

Thus,

$$V(f) = 0 \; , \tag{4.4}$$

hence also

$$\frac{d}{dt} f(x(t)) = 0 \tag{4.5}$$

along all solutions $t \to x(t)$ of (4.1), i.e., f is a constant-of-motion of the differential equations (4.1).

Now, we want to set up non-trivial partial differential equations whose solutions will determine sets of (L_a, M_a^b). We can do this in the following way. Suppose $\mathcal{A}$ is an *abelian* Lie algebra of vector fields on X such that the following condition is satisfied:

$$[A_1, [A_2, V]] = 0 \qquad (4.6)$$

for $A_1, A_2 \in \mathcal{A}$.

In the case of Hamiltonian mechanics, where $x \in X$ is the set of (q^i, p_i), the $\mathcal{A}$ can be taken as the Lie algebra of vector fields generated by the vector fields

$$\frac{\partial}{\partial p_i}.$$

Notice that the $\mathcal{A}$, in this case, generates what is now called a *Lagrangian foliation*.

In view of condition (4.6), it is natural to look for (L_a, M_a^b), which also satisfy (4.6), i.e.,

$$A_1 A_2 (L_a) = 0$$
$$A_1 A_2 (M_a^b) = 0 \qquad (4.7)$$

for $A_1, A_2 \in \mathcal{A}$.

$\mathcal{A}$ determines a foliation (with singularities) on X. Let

$$Q = \mathcal{A} \setminus X \qquad (4.8)$$

be the quotient space. Just as was done in Section 3, one can define bundles on Q such that L_a and M_a^b are constructed from the cross-sections of the bundles. One can see this in a coordinate free way by using condition (4.6) and applying $\text{Ad}(A_1) \text{Ad}(A_2)$ to both sides of (4.11)

$$[A_1, V](L_a) + V(A_1(L_a)) = A_1(M_a^b) L_b + M_a^b A_1(L_b) \qquad (4.9)$$

$$[A_1, V](A_2 L_a) + [A_2, V](A_1(L_a)) = A_1(M_a^b) A_2(L_b) + A_2(M_a^b) A_1(L_b). \qquad (4.10)$$

For $A \in \mathcal{A}$, set

$$L_a(A) = A(L_a) \qquad (4.11)$$

$$M_a^b(A) = A(M_a^b) \qquad (4.12)$$

$$V(A) = [A, V]. \qquad (4.13)$$

Then, we have:

Calogero Equations

$$\mathcal{A}(L_a(A)) = 0$$
$$= \mathcal{A}(M_a^b(A)) \quad . \tag{4.14}$$

(4.14) implies that the functions $L_a(A)$, $M_a^b(A)$ are *functions* on Q. Similarly,

$$[\mathcal{A}, V(A)] = 0 \quad ,$$

hence V is the pull-back of a cross-section of a vector bundle, which lies in Q. Equation (4.10) can then be interpreted (as we did in Section 3) as certain differential equations involving differential operators on vector bundles which live on Q.

References

1. M. Toda, *Theory of Nonlinear Lattices*, Springer-Verlag, 1981.

2. E.T. Whittaker, *A Treatise on Analytical Dynamics of Particles and Rigid Bodies*, Cambridge Univ. Press, London and New York, 1959.

3. H. Flaschka, *Phys. Rev. B* **9**, 1924 (1974).

4. P.D. Lax, *Comm. Pure Appl. Math.* **21**, 487 (1968).

5. F. Calogero, *Lett. Nuovo Cimento* **13**, 44 (1975); **16**, 77 (1976).

6. R. Hermann, The geometric foundation of the integrability property of differential equations and physical systems, Parts I and II, submitted to *J. Math. Phys.*

7. R. Hermann, *Vector Bundles in Mathematical Physics*, Vol. I, W.A. Benjamin, Reading, MA, 1970.

8. E. Goursat, *Lecons sur l'intégration des équations aux derivés partielles du second ordre*, Vols. I and II, Gauthier-Villars, Paris, 1890.

9. E. von Weber, G. Floquet, and E. Goursat, Proprietés générales des systems d'equations aux derivées partielle, *Encyclopédie des Sciences Mathématiques*, Tome II, Vol. 4, Gauthier-Villars, Paris, 1913.

10. C. Riquier, *Les Systèmes D'Equations aux Derivées Partielles*, Gauthier-Villars, Paris, 1910.

11. A. Forsyth, *Theory of Differential Equations*, Dover, New York.

Chapter 4

EHRESMANN PSEUDOGROUPS AND FOLIATIONS

1. EHRESMANN PSEUDOGROUPS ON A FIXED MANIFOLD

In the 1950's, C. Ehresmann developed a major foundational concept for differential geometry, the *pseudogroup*. Unfortunately, he presented his ideas (which were abstracted from the classical work of Lie, Vessiot, and Cartan) only in short notes and conference proceedings, hence his ideas and theories have not penetrated directly to the physics and applied mathematics world. In this section I will recapitulate some of this material in a form that will be useful in the broad areas of physics and system theory, partially following ideas of Plante [2]. A key concept in Ehresmann's work is the generalization of the "linear isotropy subgroup of a transformation group" concept to a pseudogroup and the *application* to foliation theory. R. Bott [3] later showed that, for the case of a pseudogroup attached to a nonsingular foliation, this holomony group was the holomony group attached to a *flat* affine connection on the leaves. He deduced several interesting global consequences of this fact, which I will not go into here. In [4,5] I developed relations between the Ehresmann holomony group and geometric structures defined on the quotient and fibers of a nonsingular foliation.

This work was done with no applications in mind. However, in recent years applied disciplines have arisen which involve, indirectly, Ehresmann pseudogroups, foliation and holomony ideas. I will cite two examples:

> The Krener-Isodori theory of disturbance decoupling for nonlinear input-output systems.

> The theory of integrable physical systems which involve ordinary differential equations (possibly on infinite dimensional manifolds), which have associated *symmetries*. These symmetries define the pseudogroups, the isotropy-holomony of this pseudogroup plays a key role in the theory of integrable systems.

<u>Definition</u>. Let X be a C^∞, paracompact manifold. A *local* C^∞ *map* is a triple

$$\delta = (D, \phi, R)$$

consisting of open subsets D, R of X, and a (C^∞) map $\phi: D \to R$ mapping D *onto* R. D is the *domain* of δ, R the *range*. If ϕ is a *diffeomorphism* between D and R, we say that δ is a *local diffeomorphism* for X.

If $\delta = (D,\phi,R)$ is such a local map, with ϕ a diffeomorphism, define:

$$\delta^{-1} = (R, \phi^{-1}, R) \, , \qquad (1.1)$$

where ϕ^{-1} is the inverse map to ϕ, a diffeomorphism from R to D.

If

$$\delta = (D,\phi,R) \, ,$$

$$\delta' = (D',\phi',R')$$

are local maps, we say that

$$\delta \subset \delta' \qquad (1.2)$$

if the following conditions are satisfied:

$$R \subset R'$$
$$D \subset D' \qquad (1.3)$$
$$\delta = \delta' \text{ restricted to } D \, .$$

If $\delta = (D,\phi,R)$, $\delta' = (D',\phi',R')$ are local maps, we say that

$$\delta \sim \delta' \qquad (1.4)$$

if the following condition is satisfied:

$$\phi(x) = \phi'(x) \qquad (1.5)$$
$$\text{for all } x \in D \cap D'$$

If δ, δ' satisfy (1.4), we will define

$$\delta \cup \delta' = (D'', \phi'', R'') \qquad (1.6)$$

as follows:

$$D'' = D \cup D'$$
$$R'' = R \cup R'' \qquad (1.7)$$

$$\phi''(x) = \phi(x) = \phi'(x)$$
$$\text{for } x \in D \cap D'$$

$$\phi''(x) = \phi(x)$$
$$\text{for } x \in D$$

$\phi''(x) = \phi'(x)$

for $x \in D'$.

It is readily seen that these formulae define ϕ'' as a C^∞ map. However, of course it may not be a diffeomorphism, even if ϕ and ϕ' are diffeomorphisms.

Suppose $\delta = (D,\phi,R)$, $\delta'(D',\phi',R')$ are local diffeomorphisms. Define another local diffeomorphism

$$\delta \circ \delta' = (D'',\phi'',R'') \tag{1.8}$$

as follows:

$$D'' = {\phi'}^{-1}(D \cap R') \tag{1.9}$$

$$R'' = \phi(D \cap R') \tag{1.10}$$

$$\phi''(x) = \phi(\phi'(x)) \tag{1.11}$$

for $x \in D''$.

<u>Definition</u>. A collection Δ of local diffeomorphisms of the manifold X is said to be an *Ehresmann pseudogroup acting on* X if the following conditions are satisfied:

1) The identity map

$$1 = (X,id,X)$$

is an element of Δ.

2) If $\delta \in \Delta$, then $\delta^{-1} \in \Delta$.

3) If $\delta,\delta' \in \Delta$, then

$$\delta \circ \delta' \in \Delta .$$

4) If $\delta' \in \Delta$, and $\delta \subset \delta'$, then $\delta \in \Delta$

5) If $\delta,\delta' \in \Delta$ with $\delta \sim \delta'$, and if $\delta \cup \delta'$ is a local diffeomorphism, then $\delta \cup \delta' \in \Delta$.

2. PSEUDOGROUPS AS THE AUTOMORPHISMS OF TENSOR FIELDS

The most important pseudogroups in both geometry and physics are those defined via tensor fields. A *tensor field* τ on a manifold X is a smooth cross-section of a tensor product bundle of the tangent bundle $T(X)$ and its dual $T^d(X)$. (Alternately, it may be considered as an associated bundle of the principal tangent bundle with structure group $GL(n,R)$, defined by a linear action of $GL(n,R)$ on a real vector space.)

A local diffeomorphism $\delta = (D,\phi,R)$ is then said to be a *symmetry* (or automorphism) of τ if, for each $x \in D$, the linear map $\phi_*: X_x \to X_{\phi(x)}$, acting on tensors, sends the value of τ at x to the value of τ at (x). It is readily seen that the set of such symmetries of τ defines an Ehresmann pseudogroup, as defined above. We will call such a pseudogroup a *tensorial pseudogroup*. It is a special case of a *Lie pseudogroup* [6].

The pseudogroups encountered in physics and geometry often have a special property, that one calls "flatness".

<u>Definition</u>. A tensor field on a manifold X is said to be *locally flat* if the following condition is satisfied:

Each point x has a neighborhood U and a coordinate system (x^i) defined on U such that the components of τ in U with respect to the bases of tensor spaces constructed from $\partial/\partial x^i$ and dx^i (which are cross-sections of $T(U)$ and $T^d(U)$) are *constant* in U.

3. EXAMPLES OF LOCALLY FLAT TENSOR FIELDS AND THEIR ASSOCIATED PSEUDOGROUPS SYMPLECTIC STRUCTURES

Consider first a two-form ω on X such that

$$d\omega = 0 \qquad (3.1)$$

The space of Cauchy characteristic vectors

$$\mathscr{C}(\omega)(x) = \{v \in X_x : v \lrcorner \omega = 0\} \qquad (3.2)$$

is constant as x varies over X.

By the classic Darboux theorem, every point $x \in X$ has a neighborhood with a set

$$\{q^i, p_i\}, \qquad 1 \leq i, j \leq n$$

of functions such that:

$$\omega = dp_i \wedge dq^i$$

$$\omega^n = dp_1 \wedge \cdots \wedge dp_n \wedge dq^1 \wedge \cdots \wedge dq^n \neq 0$$

The Cauchy characteristic vectors $\mathscr{C}(\omega)$ define a foliation on X. The quotient space

$\mathscr{C}(\omega) \setminus X$

by this foliation admits a *symplectic structure* in the sense that is now standard in classical and quantum mechanics. The symmetry pseudogroup of this tensor field is a basic geometric object for this application

Complex Structures

Let X be an even dimensional manifold. A *complex structure* on X is defined by an $\mathscr{F}(X)$ linear map

$$J: \mathscr{V}(X) \to \mathscr{V}(X)$$

such that the following linear subspace:

$$\{V + iJ(V) : V \in \mathscr{V}(X)\}$$

of $\mathscr{V}(X) \otimes \mathbb{C}$ is a Lie subalgebra of the Lie algebra of all complex-valued, C^∞ vector fields on X.

The *Newlander-Neurenberg theorem* [ ] then asserts that J is locally flat, i.e., each point $x \in X$ has a coordinate neighborhood U with a coordinate system

$$(x^i, y^j)$$
$$1 \le i, j \le n$$
$$2n = \dim X$$

such that:

$$J\left(\frac{\partial}{\partial x^i}\right) = \frac{\partial}{\partial y^i}$$

$$J\left(\frac{\partial}{\partial y^i}\right) = -\frac{\partial}{\partial x^i}$$

Locally Flat Riemann Manifolds

Let

$$g: \mathscr{V}(X) \times \mathscr{V}(X) \to \mathscr{F}(X)$$

be a nondegenerate symmetric, $\mathscr{F}(X)$-bilinear form on vector fields, i.e., g defines a *Riemann metric*. In terms of coordinates (x^i) and tensor analysis notation,

$$g = g'_{ij} \, dx^i \, dx^j$$

i.e.,

$$g\left(\frac{\partial}{\partial x^i}, \frac{\partial}{\partial x^j}\right) = g^{ij}.$$

The classic condition (denoted by Riemann as   ) that this tensor field be locally flat is that the *Riemann curvature* tensor be zero.

Various "singular" versions of Riemannian metrics can be defined and play an important (but largely unrecognized) role in applied mathematics and mathematical physics. For example,

$$\Delta = g^{ij}\frac{\partial}{\partial x^i \partial x^j} + \frac{\partial}{\partial x^i} + h$$

is a linear, second-order, partial differential operator, then its *symbol*

$$\sigma(\Delta) = g^{ij}\frac{\partial}{\partial x^i} \otimes \frac{\partial}{\partial x^j}$$

is a *contravariant* tensor field, i.e., what I have called a *co-Riemannian metric*. The condition for local flatness of these tensor fields is not known. The theory of *Riemannian submersions* originated by B. Reinhart [8] also involves the co-Riemannian structures.

4. PSEUDOGROUPS GENERATED BY LIE ALGEBRAS OF VECTOR FIELDS

We can also consider relations between Ehresmann pseudogroups and sets of vector fields on X. Let V be such a vector field. Let $t \to \exp(tV)$ be the one-parameter pseudogroup it generates, i.e., the collection of local diffeomorphisms: $D \to R$ obtained by finding the orbit curves of V, starting at points x D, then going out t time units.

<u>Definition</u>. V is associated with the Ehresmann pseudogroup if each local diffeomorphism generated by the orbit curves of V belongs to D.

Standard results prove the following:

<u>Theorem 4.1</u>. Let τ be a tensor field on a manifold X and let $\mathcal{G}(\tau)$ be the Lie algebra of vector fields $V \in \mathcal{V}(X)$ such that

$$\mathcal{L}_V(\tau) = 0 \qquad (4.1)$$

($\mathcal{L}$ denotes "Lie derivative".) Then, the pseudogroup of symmetries of τ contains the one-parameter pseudogroup generated by the vector fields in $\mathcal{G}(\tau)$.

Now, for the "linear isotropy subgroup" concept for a pseudogroup.

<u>Definition</u>. Let $\mathscr{P} = (D,\phi,R)$ be a pseudogroup on a manifold X. Let x be a point of X. Let $(D,\phi,R) \in \mathscr{P}$ be the set of elements of pseudogroups such that the following conditions are satisfied:

$$x \in D \cap R$$
$$\phi(x) = x \; . \tag{4.2}$$

Associate to (D,ϕ,R) the linear map

$$\phi_*: X_x \to X_x \; .$$

The collection of linear maps on the tangent space X_x --which, it is readily seen, forms a group called the *linear isotropy group of the pseudogroup* $\mathscr{P}$ *at* x.

This concept unifies many special situations encountered in classical and modern differential geometry and differential equation theory, as we shall now describe.

Given a Lie algebra $\mathscr{G}$ of vector fields, we can form the *pseudogroup it generates*, i.e., the smallest pseudogroup containing the

$$\exp(tV)$$
$$t \in R, \quad V \in \mathscr{G}$$

Of course, a particular case is that where G is a Lie group which acts on X,

$$G \times X \to X$$
$$(g,x) \to gx \; ,$$

with $\mathscr{G}$ the Lie algebra of G, considered as a Lie algebra of vector fields on X, as the *infinitesimal version* of the action of G on X. In this case, the following facts are standard, and follow from the elementary parts of calculus on manifolds:

The orbits

$$Gx \equiv \{gx: g \in G\}$$

are submanifolds. If

$$G^x = \{g: gx = x\}$$

is the isotropy subgroup of G at x, the orbit G_*

is isomorphic to G/G^x. G^x acts on X_x: g applied to $v \in X_x$ is the result of g_* applied to v.

This is called the *linear isotropy group action* of the transformation group.

5. THE LINEARIZATION OF A VECTOR FIELD WITH A FIXED POINT

A traditional situation (at least since Poincaré) in the theory of ordinary differential equations is the study of solutions of a nonlinear ordinary differential equation in R^n:

$$\frac{dx}{dt} = f(x) \tag{5.1}$$

$$x \in R^n$$

in the neighborhood of a point, which we shall take to be $x = 0$, at which the right hand side of (5.1) vanishes:

$$f(0) = 0 .$$

The solutions of (5.1) are--by the very basis of Lie theory--*orbit curves* of the vector field

$$V = f(x) \frac{\partial}{\partial x} . \tag{5.2}$$

Theorem 5.1. Identify the tangent space to R^n at $x = 0$ with R^n itself. Write the Taylor series of $X \to f(x)$ at $x = 0$

$$f(x) = Ax + \cdots \tag{5.3}$$

where A is an $n \times n$ matrix. Consider the one-parameter pseudogroup generated by V,

$$t \to \exp(tV)$$

and the action on $R_0 \equiv R^n$. Then, the planar isotropy group of the pseudogroup is the one-parameter linear group

$$t \to \exp(tA) .$$

Proof. To determine the linear isotropy group at $x = 0$, we must consider the image under $\exp(tV)$ of curves

$$r \to x(s)$$

which begin at $= 0$, i.e., which are such that

$$x(0) = 0 \ .$$

Set:

$$x(t,s) = \exp(tV)(x(s)) \ .$$

Then,

$$\frac{\partial x}{\partial t} = f(x(t,s))$$

$$= Ax(t,s) + \cdots$$

$$x(0,s) = x(s) \ .$$

To determine how $\exp(tV)$ transforms the tangent vector

$$\frac{dx}{ds}(0) \equiv \frac{\partial x}{\partial s}(0,0)$$

we must then calculate

$$\frac{\partial x}{\partial s}(t,0) \ ,$$

i.e.,

$$\exp(tV)(0) = 0$$

$$\frac{\partial x}{\partial s}(t,0) = \exp(tV)_*\left(\frac{dx}{ds}(0)\right)$$

But

$$\frac{\partial}{\partial t}\frac{\partial x}{\partial s}(t,0) \quad \frac{\partial^2 x}{\partial s \partial t}(t,0) = f_x(x(t,0))\frac{\partial x}{\partial s}(t,0)$$

$$= f_x(0)\frac{\partial x}{\partial s}(t,0)$$

$$= A\frac{\partial x}{\partial s}(t,0)$$

Then

$$\frac{\partial x}{\partial s}(t,0) = \exp(tA)\left(\frac{\partial x}{\partial s}(0,0)\right) \ ,$$

which proves Theorem 5.1.

6. THE POINCARE MAP ASSOCIATED WITH A PERIODIC ORBIT AS THE LINEAR ISOTROPY SUBGROUP OF A ONE-PARAMETER PSEUDOGROUP

Let us now consider the other prototype of Ehresmann's basic idea in the classical theory of nonlinear ordinary differential equations. As in Section 5, consider an ordinary differential equation on R^n, of the following form:

$$\frac{dx}{dt} = f(x) . \qquad (6.1)$$

Consider a solution

$$t \to x_0(t)$$

of (6.1), which is *periodic* of period $T > 0$, i.e.,

$$x(0) = 0$$
$$x_0(t+T) = x_0(t) \qquad (6.2)$$
$$x_0(t) \ne x_0(0)$$

for $0 < t < T$.

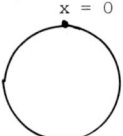

$x = 0$

Let $t \to \exp(tV)$, $V = f(x)(\partial/\partial x)$ again be the pseudogroups generated by the vector field V.

As in Section 5, we can consider the *linear variational equations* of (6.1) associated with the periodic solution $t \to x_0(t)$:

$$\frac{\partial \hat{x}}{\partial t} = f_x(x_0(t)) \frac{\partial \hat{x}}{\partial t} . \qquad (6.3)$$

Consider the pseudogroup $t \to \exp(tV)$ and, for fixed t, the local diffeomorphism

$$(D, \phi, R)$$

it generates. Suppose

$$\phi(0) = 0 .$$

By the uniqueness theorem for solutions of (6.1), $\phi(x_0)$, for x sufficiently

Ehresmann Pseudogroups 69

close to 0, must result from solving (6.1), with initial conditions at x. We see then that the isotropy subgroup of the pseudogroup is the diffeomorphism

$$\exp(ntV)$$

$$n = 0, \pm 1, \ldots$$

As in the proof of Theorem 5.1, the action of this on tangent vectors is the result of solving the *Poincaré linearization (6.3)*.

7. PSEUDOGROUPS AND HOLONOMY GROUPS ASSOCIATED WITH FOLIATIONS

Let X continue as a finite dimensional, C^∞, paracompact manifold. Let $\mathscr{W}$ be an $\mathscr{F}(X)$-submodule of $\mathscr{V}(X)$ such that:

$$\mathscr{F}(X) \mathscr{W} \subset \mathscr{W} \qquad (7.1)$$

i.e., $\mathscr{W}$ is an $\mathscr{F}(X)$-submodule of $\mathscr{V}(X)$, and

$$[\mathscr{W}, \mathscr{W}] \subset \mathscr{W} \qquad (7.2)$$

i.e., $\mathscr{W}$ is a Lie subalgebra. For $x \in X$, set:

$$\mathscr{W}(x) = \{w(x): x \in \mathscr{W}\}$$

A precise smooth curve

$$t \to x(t) , \qquad a \leq t \leq b$$

on X is an *orbit curve* of $\mathscr{W}$ if the following condition is satisfied:

$$\frac{dx}{dt} \in \mathscr{W}(x(t)) \qquad (7.3)$$

for all $t \in [a,b]$.

<u>Definition</u>. $\mathscr{W}$ is said to define a *foliation* for X if (7.1) and (7.2) are satisfied, plus the following condition:

$$\dim \mathscr{W}(x(t)) \text{ is a constant for } a \leq t \leq b \qquad (7.4)$$
$$\text{where } t \to x(t) \text{ is any orbit curve of } \mathscr{W}.$$

<u>Theorem 7.1</u>. (Frobenius-Chevalley, as extended to the singular case in [5].) Suppose $\mathscr{W}$ defines a foliation, in the sense defined above. Then, every point $x \in X$ is constant in a unique immersed submanifold $Y(x) \subset X$ such that

$$Y(x)y = \mathscr{W}(y)$$

for $y \in Y(x)$

$Y(x)$ contains all orbit curves of $\mathcal{W}$ which begin at x.

The decomposition $x \to Y(x)$ of X into submanifolds defines an equivalence relation on X; the quotient space is denoted as

$$\mathcal{W} \setminus X \quad .$$

Each submanifold $Y(x)$ is called a *leaf* of the foliation.

<u>Proof</u>. Given in [5], an exterior of the original for the nonsingular case given by Chevalley in his proof of the global version of the Frobenius theorem in his "Lie groups" treatise.

<u>Remark</u>. Condition (7.4) is not satisfactory, because it is non-constructive. It is implied by a number of other conditions which arise naturally in geometric and applied situations:

a) $\quad \mathcal{W}$ is nonsingular, in the sense that:

$$\dim(\mathcal{W}(x)) \text{ is constant, for } x \in X.$$

b) $\quad X$ and $\mathcal{W}$ are *real analytic*.

c) $\quad \mathcal{W}$ is *locally finitely generated* in the sense defined in [5].

Suppose $\mathcal{W}$ defines a foliation, in the above sense. We can then define the *pseudogroup and the foliation* as the smallest pseudogroup on X containing the diffeomorphism

$$\exp(V) \quad ,$$

where $V \in \mathcal{W}$. The *holomony of the foliation* at the point Y is the group of linear transformation on the normal vector space

$$X_x / Y(x)_x \equiv Y(x)_x^\perp$$

to the leaf $Y(x)$ generated by the *linear isotropy group* of the foliation pseudogroup.

Ehresmann gave it the name "holomony group", and presumably took it as obvious that there was a connection in the normal vector bundle

$$Y(x)^\perp = \{(y,v): Y \in Y(x); v \in X_y/Y(x)_y\} \tag{7.5}$$

to the leaf $Y(x)$ whose holomony group, in the connection sense, is the group

Ehresmann Pseudogroups 71

defined above using pseudogroup ideas. R. Bott seems to have been the first one to have provided this connection explicitly.

We shall define both connections as a covariant derivative operator

$$(V',\gamma) \to \nabla_{V'}\gamma$$

$V' \in \mathscr{V}(Y) \equiv$ vector fields on leaf Y

$\gamma \in$ cross-sections of the normal vector bundle $Y^\perp$

in the following way:

For vector fields V_1, V_2 with:

$$V_1 \in \mathscr{W} \quad , \quad V_2 \in \mathscr{V}(X)$$

consider γ_2, the cross-section of $Y^\perp$, defined as follows:

$$\gamma_2(y) = [V_1, V_2](y) \text{ projected modulo } Y_y = \mathscr{W}(y) \qquad (7.6)$$

for all $y \in Y$.

When fV_1 is substituted for V_1, for $f \in \mathscr{F}(X)$, note that:

$$[fV_1, V_2] = f[V_1, V_2] - V_2(f)V_1 ,$$

the second, non-tensorial, term on the right hand side of (7.6) vanishes when projected modulo $\mathscr{W}$.

Thus, if we choose V_1 so that:

$$V_1 = V' \text{ restricted to } Y ,$$

and V_2 so that:

$$\gamma(y) \equiv V_2(y) \text{ projected modulo } \mathscr{W}(y) ,$$

and define

$$\nabla_{V'}\gamma = \gamma_2 \qquad (7.7)$$

we see that the resulting operation is, first, *well defined*, and, second, satisfies the conditions to define a linear connection (in the Kozul sense) for the normal vector bundle $Y^\perp$ to the submanifold Y.

<u>Theorem 7.2</u>. The holomony group for the Bott connection, in the usual sense of connection theory, is the holomony group as defined by Ehresmann using the pseudogroup generated by $\mathcal{W}$.

<u>Proof</u>. This follows readily after making explicit the differential equations which define parallel transport for this connection. (7.7) shows that these equations essentially involve the Jacobi bracket structure determined by the pseudogroup.

<u>Theorem 7.3</u>. If the foliation $\mathcal{W}$ is nonsingular, then the Bott connection ∇ for the normal vector bundle $Y^\perp$ to each leaf Y has curvature zero.

<u>Proof</u>. To prove the curvature is zero it suffices to prove that there are, locally, m vector fields

$$V_1, \ldots, V_m ,$$

$$m = \dim X - \dim Y$$

such that

$$\nabla_V V_1 = 0 = \cdots = \nabla_V V_m$$

for all $V \in \mathcal{V}(Y)$,

i.e., the $V_1, \ldots, V_m$ are *flat*. In view of (7.6), this means that

$$[\mathcal{W}, V_i] \subset \mathcal{W}$$

$$i = 1, \ldots, m .$$

That such vector fields exist locally is a consequence of the Frobenius complete integrability theorem.

References

1. C. Ehresmann, Sur les structures infinitesimals regulieres, *Congr. Int. Math. Amseterdam* <u>1</u>, 479-480 (1954); Connexions infinitesimales, *Coll. Top. Alg. Bruxelles*, 29-55 (1950); Structures infinitesimales et pseudogroupes de Lie, *Colloq. Int. C.N.R.S. Geom. Diff. Strasbourge*, 97-110 (1953); *C.R. Acad. Sci. Paris* <u>240</u>, 1762 (1954); <u>241</u>, 397, 1755 (1955); <u>246</u>, 360 (1958); Connexions d'ordre superieur, *Atti 5 Congresso dell' Unione Matematica Italiana 1955*, Cremonese, Rome, 1956, pp. 326-8; Categories topologiques et categories differentiables, *Colloq. Geom. Diff. Globale Bruxelles*, *C.B.R.M.*, 137-150 (1958); Groupoides differenciales, *Rev. In. Mat. Argentina, Buenos-Aires* <u>19</u>, 48 (1960).

2. J. Plante, Foliations with measure preserving holonomy, *Ann. Math.* <u>102</u>, 327-61 (1975).

3. R. Bott, On a topological obstruction to integrability. In *Global Analysis, Proceedings of Symposia in Pure Math.*, 16, 127-131, Providence RI, Amer. Math. Soc, 1970.

4. R. Hermann, The differential geometry of foliations 1, *Ann. of Math. (2)* 72, 445-457.

5. R. Hermann, Cartan connections and the equivalence problem for geometric structures, *Cartan Diff. Eq.* 3, 199-248 (1964).

6. A. Kumpera and D.C. Spencer, *Lie Equations*, Princeton Univ. Press, 1972. See Bibliography for further references to extensive work of the Spencer school.

7. A. Newlander and L. Nirenberg, Complex analytic coordinates in almost complex manifolds, *Ann. Math.* 65, 391-404 (1957).

8. B. Reinhart, Foliated manifolds with bundle-like metrics, *Ann. Math.* 69, 119-132 (1959).

Chapter 5

LINEAR DIFFERENTIAL OPERATORS AND
PARTIAL WAVE ANALYSIS

1. INTRODUCTION

The study of linear, variable-coefficient differential operators and equations is, of course, fundamental for all of mathematical physics. There has recently been extensive development of the mathematical theory from the point of view of functional analysis -- such topics as *pseudo-differential operators* and *Fourier integral operators* are hybrids of traditional mathematical physics lore and the new insights that have resulted from the influx of functional analysis techniques.

Now, the classical mathematical physics point of view in the theory of linear differential equations leads in two directions -- of course, toward the functional analytic, but also toward the less precisely defined areas of differential geometry (fiber bundles, connections, Riemannian metrics, deformations, etc.) and Lie theory. My aim is to indicate briefly certain viewpoints.

2. GENERALITIES ABOUT LIE TRANSFORMATION GROUPS, LINEAR DIFFERENTIAL OPERATORS ON MANIFOLDS, AND ORBIT SPACES

In this section we study linear differential operators with scalar coefficients which are invariant under Lie group actions. All data will be of differentiability class C^∞, and all manifolds Hausdorff and paracompact, unless mentioned otherwise. Let X be such a manifold. $\mathscr{F}(X)$ denotes the C^∞ real-valued functions on X. $\underline{D}(X)$ denotes the linear differential operators: $\mathscr{F}(X) \to \mathscr{F}(X)$. (They can either be defined "classically" in terms of local coordinates, or algebraically, e.g., as in *Geometry, Physics and Systems*. $\underline{D}(X)$ is an associated algebra (the usual product of operators) and filtered

$$\underline{D}^0(X) = \mathscr{F}(X) \subset \underline{D}^1(X) \subset \cdots$$

with $\underline{D}^n(X)$ the operator of order n in the usual sense. $\underline{D}(X)$ is also a Lie algebra with Lie algebra operation

$$(D_1, D_2) \to [D_1, D_2] \equiv D_1 D_2 - D_2 D_1$$

the operator commutator.

Let DIFF(X) be the group of diffeomorphisms of X. Let G be a group. A *transformation group action* is a homomorphism

$$G \to \text{DIFF}(X) \;.$$

Since we will usually only be dealing with one such homomorphism, the notation will be simplified by denoting as gx the transform of $x \in X$ by the diffeomorphism assigned to $g \in G$. We obtain in this way a map

$$G \times X \to X$$

$$(g,x) \to gx \;.$$

If G is a *Lie group*, and this map is C^∞, we call it a *Lie transformation group action*.

$\mathscr{V}(X)$, the vector fields on X, we defined as the *derivations* of $\mathscr{F}(X)$. If $g \in G$, $f \in \mathscr{F}(X)$, let $g^*(f)$ denote the function

$$x \to f(gx)$$

on X. The action

$$(g,f) \to g^{-1*}(f)$$

defines a linear action $\mathscr{F}(X)$, which is dual to the given transformation group action.

<u>Remark</u>. In quantum mechanics, the action of G on the "states" X is the "Schrodinger picture", while the dual action on "observables" $\mathscr{F}(X)$ is the "Heisenberg picture". In this way, $\mathscr{V}(X)$ is imbedded as a Lie subalgebra of $\underline{D}^1(X)$. ($\underline{D}^1(X)$ is a Lie subalgebra of $\underline{D}(X)$.) $\underline{D}^1(X)$ is a semi-direct sum (in the Lie algebra sense) of the ideal $\mathscr{F}(X) \equiv \underline{D}^0(X)$ and the Lie subalgebra $\mathscr{V}(X)$.

If G is a Lie transformation group on X, then its Lie algebra $\mathscr{G}$ acts as a Lie subalgebra of $\mathscr{V}(X)$:

$$V(f) = \frac{d}{dt}(\exp(-tV)^*(f))\Big|_{t=0}$$

for $f \in \mathscr{F}(X)$, $V \in \mathscr{G}$

Linear Differential Operators 77

A differential operator $D \in \underline{D}(X)$ is *invariant* under the group action G if

$$g*D = Dg*$$

for all $g \in G$.

If G is a connected Lie group, which acts in a Lie way on X, this condition is equivalent to

$$[V,D] = 0$$

for all $V \in \mathcal{G} \subset \mathcal{V}(X)$.

3. DIFFERENTIAL OPERATORS INVARIANT UNDER A ONE-PARAMETER GROUP

Let $D \in \underline{D}^n(X)$ be an n-th order linear differential operator on a manifold X which is invariant under a vector field $V \in \mathcal{V}(X)$, i.e.,

$$[D,V] = 0 \tag{3.1}$$

Suppose, for the moment, that V is non-zero at each point of X, i.e., that the group generated by V has no fixed points. (We shall consider the possibility of zero points of V later on.) Let us choose a local coordinate system (x^i), $1 \leq i,j \leq n$, for X. Let $\partial_i = \partial/\partial x^i$ be the vector fields defined by these coordinates, in the open subset U of X. Suppose the coordinates are chosen such that

$$V = \partial_1 . \tag{3.2}$$

Then, in this neighborhood D can be written in the following form:

$$D = \Delta_0^U V^n + \Delta_1^U V^{n-1} + \cdots + \Delta_n^U , \tag{3.3}$$

where:

$$\Delta_0^U, \ldots, \Delta_n^U \in \underline{D}(U) . \tag{3.4}$$

$\Delta_0^U, \ldots, \Delta_n^U$ are differential operators on the variables $(x^2, \ldots, x^n)$ *only*, such that

$$[V, \Delta_0^U] = 0 = \cdots = [V, \Delta_n^U] \tag{3.5}$$

X can be covered by open subsets $\{U\}$ of this type. We obtain systems

$$\Delta^U = \{\Delta_0^U, \ldots, \Delta_n^U\} \tag{3.6}$$

of locally defined differential operators. In the intersection $U \cap U'$ of two open subsets of this system, the operators Δ^U and $\Delta^{U'}$ are related by certain linear relations, i.e., the $\{\Delta^U\}$ can be interpreted as cross-sections of certain *fiber bundles* over X. The geometric and topological properties of this bundle might be studied. My plan is to assume that representations of the form (3.3) can be found *globally* on X, and study the analytic consequences. However, let us first review certain background of the differential geometry of Lie group transformation actions.

4. REGULAR LIE TRANSFORMATION GROUPS

Let G be a Lie group acting as a transformation group on a manifold X, i.e., a (C^∞) map

$$(g,x) \to gx$$
$$G \times X \to X \tag{4.1}$$

is defined satisfying the usual relations.

Given $x \in X$, let

$$Gx = \{gx: g \in G\}$$

be the orbit of x under G. Each orbit is a submanifold of X, and the orbits form a *partition* of X, i.e., each point of X lies in *precisely* one orbit. The space of all orbits (as an abstract set) is denoted as

$$G \backslash X \ .$$

Denote the Lie algebra of G as $\mathcal{G}$. A basic fact in Lie group theory (which we shall assume to be known) is that $\mathcal{G}$ can be identified with the set of one-parameter subgroups of G. For $A \in \mathcal{G}$, let

$$t \to \exp(tA)$$

denote the one-parameter subgroup of G associated with A.

The action (4.1) of G on X gives an action of the one-parameter group $t \to \exp(tA)$ on X. The infinitesimal generator of this one-parameter group is a vector field on X. One obtains in this way a linear map $\mathcal{G} \to \mathcal{V}(X)$ which is also a Lie algebra homomorphism. If the group action is *locally effective*, i.e., the subgroup of elements $g \in G$ which act on the identity is *discrete*, then the Lie algebra homomorphism is one-one. It will be convenient for us to only deal with locally effective action, unless mentioned explicitly otherwise.

Linear Differential Operators 79

Given a point $x \in X$, set:

$$G^x = {g \in G: gx = x} .$$

G^x is called the *isotropy subgroup* of G at x. G^x acts linearly on the tangent space to X at x. This is called the *linear isotropy representation*. This linear group leaves the tangent space $(Gx)_x$ to the orbit at x invariant, hence acts on the quotient vector space

$$X_x/(Gx)_x .$$

(This quotient is called the *normal tangent vector* to the orbit.)
Let

$$\pi: X \to G\backslash X$$

be the projection map which assigns to each $x \in X$ the orbit Gx to which x belongs.

Definition. The action of G on X is said to be *regular* if $G\backslash X$ can be given as manifold structure, such that the map π is a C^∞ submersion.

Remark. Recall that a map $\pi: X \to Y$ between manifolds is a submersion of the induced linear map

$$\pi_*: T(X) \to T(Y)$$

on tangent vectors is *onto*. The fibers of submersion maps all have the same dimension. Hence, a necessary condition that a Lie group action be a submersion is that the orbits all have the same dimension. Here is another necessary condition.

Theorem 4.1. If the action of G on X is regular, then for each $x \in X$, the action of the isotropy subgroup G^x of G on the normal tangent vectors $X_x/(Gx)_x$ to the orbit, is the *identity* action.

Proof. $\pi: X \to G\backslash X$ intertwines the action of G, with G acting as the identity on $G\backslash X$, while $X_x/(Gx)_x$ is isomorphic (via π_*) to $(G\backslash X)_{\pi(x)}$. Q.E.D.

Theorem 4.2. If X has a (positive) Riemannian metric such that G is a closed subgroup of the Lie group of isometries of X, and if all the isotropy subgroups of G are principal (i.e., if they have the same dimension and the same number of components), then the action of G on X is regular.

5. THE METHOD OF FOURIER

Let G be a Lie group which acts in a regular way on a manifold X. In addition, we shall suppose, for simplicity, that the action of G is free, i.e., that the map

$$G \to Gx$$

of G into each orbit is a diffeomorphism. (There is a method, which goes back to Lie himself, for reducing more general cases to this one.)

In this section our goal is to study the linear differential operators $D \in \underline{D}(X)$ which are invariant under the action of G. To this end, let $\sigma: G \to GL(V)$ be a representation of G by automorphisms of a finite dimensional complex vector space V. Let

$$\mathcal{F}(X,\sigma)$$

be the space of C^∞ maps

$$f: X \to V$$

such that

$$f(gx) = \sigma(g) f(x) \tag{5.1}$$

for $g \in G$, $x \in X$.

In words, f *intertwines the action of* G on X *and* U.

Relation (5.1) can also be written as:

$$g^*(f) = \sigma(g) f \tag{5.2}$$

for $g \in G$.

However, if D is invariant under G, i.e., $g^*D = Dg^*$, then

$$D(\mathcal{F}(X,\sigma)) \subset \mathcal{F}(X,\sigma) \tag{5.3}$$

$$f = f_1 v_1, \ldots, f_n v_n .$$

where $v_1, \ldots, v_n$ is a basis of V and $f_1, \ldots, f_n \in \mathcal{F}(X)$, then

$$D(f) = D(f_1) v_1 + \cdots + D_n(f_n) v_n .$$

<u>Theorem 5.1</u>. There is a vector bundle E_σ over $G \setminus X$, with fiber v isomorphic to V, such that $\mathcal{F}(X,\sigma)$ is naturally identified with $\Gamma(E)$,

Linear Differential Operators 81

the space of cross-sections of E_σ, and the action D on $\mathscr{F}(X,\sigma)$ with the action of linear differential operators D_σ on $\Gamma(E_\sigma)$.

<u>Proof</u>. This is a standard construction in fiber bundle theory. Given $f \in \mathscr{F}(X,V)$ satisfying (5.1),. Let graph(f) be the map

$$x \to (x,f(x))$$

of $X \to X \times V$. This map intertwines the action of G on X and the action $(x,v) \to (gx,(g)v)$ on $X \times V$, hence passes to the quotient to define a map on orbit spaces:

$$\gamma_f : G \backslash X \to G \backslash (X \times V).$$

$G \backslash (X \times V)$ is just the vector bundle E_σ.

Thus, we have associated with the linear differential operator D a family $\{E_\sigma, D_\sigma\}$, parameterized by the linear representations of G, of vector bundles and linear differential operators. This is, in a sense, a geometric-Lie version of the classical method of Fourier for reducing "partial" to "ordinary" differential equations.

As illustration, suppose

$$X = R^2,$$

with coordinates (t,x),

$$D(f) = f_{tt} - f_{xx} - u(x)f \qquad (5.4)$$

(Subscripts denote partial derivatives.) D is invariant under the group $G \equiv R$ of translation along the t-axis. The representations σ are parameterized by the complex numbers $\mathbb{C}$.

$$g(t,x) = (t+g,x).$$

Set:

$$\sigma(g) = e^{\sigma g}$$

$$g \in T, \qquad \sigma \in \mathbb{C}.$$

Suppose

$$f \in \mathscr{F}(X,\sigma).$$

Then,

$$f_{tt} = \sigma^2 f \qquad (5.5)$$

Then,

$$D_\sigma(f) = \sigma^2 f - f_{xx} - uf \ . \qquad (5.6)$$

Then, solving $D(f) = 0$ is reduced, in a well known way, to a Sturm-Liouville eigenvalue problem.

This method provides a general setting for much of what is called "partial wave analysis" in the mathematical physics literature.

Chapter 6

LIE'S FUNCTION GROUPS AND POISSON STRUCTURES ON MANIFOLDS

1. INTRODUCTION

As was presented in [1,2], the geometric structures called "function groups" by Sophus Lie [3] play a key underlying and *geometric* role in the study of integrable systems, quantization, etc. Some of this underlying geometry can be studied in [4-16]. In this chapter I shall, following [1], present some of this basic geometric information.

2. POISSON OPERATIONS ON MANIFOLDS AND THEIR ASSOCIATED TENSOR FIELDS

Let X be an n-dimensional C^∞, paracompact manifold, and let $\mathscr{F}(X)$ denote the set of C^∞ real valued functions on X. Let

$$T(X) = \bigcup_{x \in X} X_x \qquad (2.1)$$

be the *tangent vector bundle* and let

$$T^d(X) = \bigcup_{x \in X} X_x^d \qquad (2.2)$$

be the dual bundle, called the cotangent bundle. The one-differential forms $\mathscr{D}^1(X)$ on X are the cross-sections of T^d, while the vector fields $\mathscr{V}(X)$ are the cross sections of $T(X)$. Let

$$T(X) \wedge T(X) \qquad (2.3)$$

be the exterior product of two copies of the tangent bundle. Let

$$\Gamma(T(X) \wedge T(X))$$

be its smooth cross-sections. By the well known principles of multilinear algebra, the elements of the fiber of the bundle (2.3) above a point $x \in X$ can be identified (and we shall do so) with the skew-symmetric, bilinear maps

$$\omega_x : X_x^d \times X_x^d \to R \ .$$

A C^∞ cross-section of the bundle $T(X) \wedge T(X)$ can be identified with an $\mathscr{F}(X)$-bilinear skew-symmetric map:

$$\omega : \mathscr{D}^1(X) \times \mathscr{D}^1(X) \to \mathscr{F}(X) \ . \qquad (2.4)$$

It will be called a *bivector field* on X.

Let such an ω be given. It defines a certain type of geometric structure. We will now describe in a coordinate-free way certain geometric concepts naturally attached to this structure that were treated in a tensor-analysis framework by Schouten [7] and Nijenhuis [8].

<u>Definition</u>. Let ω be given as a cross-section of the bundle (2.3) defining a map of the type indicated in (2.4). Let $\{\,,\,\}_\omega$ be the map:

$$\mathcal{F}(X) \times \mathcal{F}(X) \to \mathcal{F}(X)$$

defined as follows:

$$\{f_1, f_2\}_\omega = \omega(df_1, df_2) \qquad (2.5)$$

for $f_1, f_2 \in \mathcal{F}(X)$.

$\{\,,\,\}_\omega$ is called the *Poisson operator* associated with the bivector field ω.

<u>Remark</u>. It will be called a *Poisson bracket* only if it satisfies the Jacobi identity, i.e., if the Schouten-Nijenhuis curvature tensor (which will be defined below) is identically zero.

<u>Theorem 2.1</u>. The Poisson operation $\{\,,\,\}_\omega$ associated with $\omega \in \Gamma(T(X) \wedge T(X))$ satisfies the following identities:

$$\{f_1, f_2\}_\omega = -\{f_2, f_1\}_\omega \qquad (2.6)$$

for $f_1, f_2 \in \mathcal{F}(X)$

$$\{f_1, f_2 f_3\}_\omega = \{f_1, f_2\}_\omega f_3 + f_2 \{f_1, f_3\}_\omega \qquad (2.7)$$

for $f_1, f_2, f_3 \in \mathcal{F}(X)$.

Let us now be algebraic [11] and consider abstractly an R-bilinear operation $\{\,,\,\}$ on $\mathcal{F}(X)$ satisfying (2.6) and (2.7). For each $f \in \mathcal{F}(X)$ set:

$$V_f(f_1) = \{f, f_1\} \qquad (2.8)$$

(2.7) says that V_f is a *derivation* of $\mathcal{F}(X)$; it can then be identified with a *vector field* on the manifold X, i.e., a cross section of the tangent bundle T(X).

Lie's Function Groups

Theorem 2.2. The mapping

$$f \to V_f$$

associated with the bivector structure on X satisfying (2.6)-(2.7) is a *first order linear differential operator* from $\mathcal{F}(X)$ to $\mathcal{V}(X)$. The *symbol* of the operator is a linear bundle map

$$\sigma: T^d(X) \to T(X) \tag{2.9}$$

such that:

$$\theta_1(\sigma(\theta_2)) = -\theta_2(\sigma(\theta_1)) \tag{2.10}$$

for $\theta_1, \theta_2 \in X_x^d$, $x \in X$.

Proof. From (2.8) we have:

$$V_{\{f_1, f_2\}} = f_1 V_{f_2} + f_2 V_{f_1} \quad . \tag{2.11}$$

This derivation rule characterizes first order linear differential operators, in the treatment of Chapter 1 of [11]. As defined there, the symbol σ of this operator assigns to each $x \in X$, $\theta \in X_x^d$ an element $\sigma(\theta)$ of X_x. Thus, X_x^d is the dual vector space to the tangent vector space X_x. $\theta_1(\sigma(\theta_2))$, as it appears in relation (2.9), is the value that the one-covector θ_1 takes on the tangent vector $\sigma(\theta_2)$. The algebraic identity (2.10) now follows readily from (2.6).

Theorem 2.3. Let $\{\ ,\ \}$ be an R-bilinear map: $\mathcal{F}(X) \times \mathcal{F}(X) \to \mathcal{F}(X)$ satisfying (2.6)-(2.7). Then, there is a unique bivector field $\omega \in \Gamma(T(X) \wedge T(X))$ such that $\{\ ,\ \}$ is *associated with* ω *in the sense of formula (2.5)*.

Proof. Set:

$$\omega(\theta_1, \theta_2) = \theta_1(\sigma(\theta_2)) \tag{2.12}$$

where σ is as in (2.9), the symbol of the assignment $f \to V_f$. Formula (2.5) is now readily verified by tracing backwards from the definition.

3. THE SCHOUTEN-NIJENHUIS TENSOR ASSOCIATED WITH A BIVECTOR FIELD

Let $\{\ ,\ \}$ be a fixed bilinear differential operator: $\mathcal{F}(X) \times \mathcal{F}(X) \to \mathcal{F}(X)$ satisfying (2.6)-(2.7). For $f_1, f_2, f_3 \in \mathcal{F}(X)$, set:

$$\Omega(f_1, f_2, f_3) = \{f_1\{f_2, f_3\}\} - \{\{f_1, f_2\}, f_3\} - \{f_2\{f_1, f_3\}\} \quad (3.1)$$

Notice that Ω is a trilinear, skew-symmetric differential operator:

$$\mathscr{F}(X) \times \mathscr{F}(X) \times \mathscr{F}(X) \to \mathscr{F}(X) \ .$$

It is identically zero if and only if $\{\ ,\ \}$ defines a Lie algebra operation on $\mathscr{F}(X)$.

For $f \in \mathscr{F}(X)$, let V_f be the vector field on X defined by:

$$V_f(f_1) = \{f, f_1\} \quad (3.2)$$

for $f_1 \in \mathscr{F}(X)$.

$f \to V_f$ is then a linear mapping of the vector space $\mathscr{F}(X)$ into the Lie algebra $\mathscr{V}(X)$. We want to find the conditions that it is a homomorphism of the algebra defined by $\{\ ,\ \}$ into the Lie algebra structure. To do this, consider $f_1, f_2, f_3 \in \mathscr{F}(X)$.

$$[V_{f_1}, V_{f_2}](f_3) = V_{f_1}(V_{f_2}(f_3)) - V_{f_2}(V_{f_1}(f_3))$$

$$= V_{f_1}\{f_2, f_3\} - V_{f_2}\{f_1, f_3\}$$

$$= \{f_1, \{f_2, f_3\}\} - \{f_2, \{f_1, f_3\}\} \quad (3.3)$$

$$V_{\{f_1, f_3\}}(f_3) = \{\{f_1, f_2\}, f_3\} \ . \quad (3.4)$$

<u>Theorem 3.1</u>. Ω is identically zero if and only if the map $f \to V_f$ is a homomorphism of the algebra structure $\{\ ,\ \}$ defined on $\mathscr{F}(X)$ into the Lie algebra structure on $\mathscr{V}(X)$.

<u>Proof</u>. Notice that (3.3) and (3.4) give the formula:

$$\Omega(f_1, f_2, f_3) = [V_{f_1}, V_{f_2}](f_3) - V_{\{f_1, f_2\}}(f_3) \ . \quad (3.5)$$

This formula makes Theorem 3.1 evident.

<u>Theorem 3.2</u>. Ω is a skew-symmetric, first order, homogeneous, trilinear differential operator. For fixed f_1, f_2 the map $f_3 \to \Omega(f_1, f_2, f_3)$ of $\mathscr{F}(X) \to \mathscr{F}(X)$ is a derivation of the associative (i.e., pointwise-product) algebra structure on $\mathscr{F}(X)$.

Proof. This is also obvious from formula (3.5).

From general principles of the algebra of multilinear differential operators (again, refer to Chapter 1 of [32]) one can now define the *symbol* of Ω, $\sigma(\Omega)$. For $x \in X$, it is a trilinear, skew-symmetric map

$$\sigma(\Omega)_x: X_x^d \times X_x^d \times X_x^d \to R \quad . \tag{3.6}$$

$\sigma(\Omega)_x$ is defined as follows. For $\theta_1, \theta_2, \theta_3 \in X_x^d$, choose $f_1, f_2, f_3 \in \mathscr{F}(X)$ such that:

$$f_1(x) = f_2(x) = f_3(x) = 0 \tag{3.7}$$

$$df_1(x) = \theta_1, \quad df_2(x) = \theta_2, \quad df_3(x) = \theta_3 \quad . \tag{3.8}$$

Then,

$$\sigma(\Omega)_x(\theta_1, \theta_2, \theta_3) = \Omega(f_1, f_2, f_3)(x) \quad . \tag{3.9}$$

As x varies, $x \to \sigma(x)_x$ defines a tensor field on X, i.e., a cross-section of the vector bundle

$$T(X) \wedge T(X) \wedge T(X) \quad .$$

This tensor field is called the *Schouten tensor* of the bivector field .

We can sum up what we have proved as follows:

Theorem 3.3. Let $\omega \in \Gamma(T(X) \wedge T(X))$ be a bivector field on the manifold X. Then, its Schouten tensor $\Omega \in \Gamma(T(X) \wedge T(X) \wedge T(X))$ is zero if and only if the associated Poisson operator $\{\ ,\ \}_\omega$ on $\mathscr{F}(X)$ satisfies the Jacobi identity, i.e., defines a Lie algebra structure on $\mathscr{F}(X)$.

Theorem 3.4. Let ω be a bivector field. If each point x of X is contained in a coordinate system such that the components of ω in this coordinate system are *constant*, then the Schouten tensor Ω vanishes. In particular, the associated Poisson structure satisfies the Jacobi identity.

Proof. This follows from the "tensorial" property of Ω.

4. PSEUDOGROUPS AND POISSON TENSORS

We can apply Ehresmann pseudogroup theory [5,12] to the study of Poisson tensors ω, and the associated Poisson operators $\{\ ,\ \}$. Let ω be a bivector field, an element of $\Gamma(T(X) \wedge T(X))$, $\{\ ,\ \}$ is an R-bilinear map $\mathscr{F}(X) \times \mathscr{F}(X) \to \mathscr{F}(X)$. The *Schouten tensor* of ω, Ω, is an element of $\Gamma(T(X)\ T(X)\ T(X))$. Let $\mathscr{G}(\omega)$ be the Lie algebra of vector fields

$V \in \mathscr{V}(X)$ such that:

$$\mathscr{L}_V(\omega) = 0 , \tag{4.1}$$

i.e.,

$$V(\omega(\theta_1, \theta_2)) = \omega(\mathscr{L}_V(\theta_1), \theta_2)) + (\theta_1, \mathscr{L}_V(\theta_2)) \tag{4.2}$$

$$\text{for all } \theta_1, \theta_2 \in \mathscr{D}^1(X) .$$

We will state a few typical results, which are readily proved using the general geometric principles proved above.

Theorem 4.1. If ω is flat, then $\Omega = 0$, and $\{ , \}$ satisfies the Jacobi identity, and makes $\mathscr{F}(X)$ into a Lie algebra.

Theorem 4.2. A vector field V is a symmetry of the bivector field if and only if it satisfies the following condition:

$$V(f_1, f_2) = V(f_1), f_2 + f_1, V(f_2) \tag{4.3}$$

$$\text{for } f_1, f_2 \in \mathscr{F}(X) ,$$

i.e., Lie derivation by V is a *derivation* of the algebraic operation on $\mathscr{F}(X)$.

Proof. That (4.3) follows from (4.2) is a routine derivation, left to the reader.

For the converse, suppose that (4.3) is satisfied. Note this proves that

$$\mathscr{L}_V(\omega)(df_1, df_2) = 0$$

$$\text{for all } f_1, f_2 \in \mathscr{F}(X) .$$

But that $\mathscr{L}_V(\omega)$ vanishes follows from the *tensorial* property of $\mathscr{L}_V(\omega)$.

Theorem 4.3. If $\Omega = 0$, if $f \in \mathscr{F}(X)$, if V_f is the vector field defined by $V_f(f') = \{f, f'\}$, then V_f satisfies:

$$\mathscr{L}_{V_f}(\omega) = 0 .$$

In particular, the one-parameter pseudogroup $t \to \exp(tV_f)$ belongs to the pseudogroup of all symmetries of ω.

5. SINGULAR FOLIATIONS AND THE FROBENIUS INTEGRABILITY THEOREM: IMPLICATIONS FOR THE POISSON STRUCTURE

I will now recapitulate work done in the context of the geometric theory of nonlinear input systems [13-18]. Let X be a manifold and let $\mathcal{W}$ be a linear subspace of $\mathcal{V}(X)$. For $x \in X$, let $\mathcal{W}(x)$ be the linear subspace $\{V(x): V \in \mathcal{W}\}$ of values of $\mathcal{W}$ at x. A continuous, piecewise C^∞ curve $t \to x(t)$, $a \leq t \leq b$, is said to be an *orbit curve* of $\mathcal{W}$ if the following conditions are satisfied:

$$\frac{dx}{dt}(t) \in \mathcal{W}(x(t)) \tag{5.1}$$

for $a \leq t \leq b$.

(dx/dt denotes the tangent vector field to the curve.) For $x_0 \in X$, let $C(\mathcal{W}, x_0)$ be the *accessible set* from x_0 along orbit curves of $\mathcal{W}$, with a set of points of X which can be joined to x_0 by a continuous, piecewise C^∞ orbit curve.

Theorem 5.1. The subsets $C(\mathcal{W}, x_0)$ of X, as x runs through X, define an equivalence relation on X for which $C(\mathcal{W}, x)$ are the equivalence classes. Each such set can be given the structure of an immersed submanifold.

The proof of this definitive form of the theorem was given by Stafan and Sussman [17,18] following the ideas of [15,16].

Suppose now that $\mathcal{W}$ is a Lie subalgebra of $\mathcal{V}(X)$, i.e.,

$$[\mathcal{W}, \mathcal{W}] \subset \mathcal{W}. \tag{5.2}$$

In general, one does not know that the accessible submanifolds $C(\mathcal{W}, x)$ are integral submanifolds of the tangent vector distribution $x \to \mathcal{W}(x)$. The following result was proved in [16].

Theorem 5.2. Suppose that (5.1) and an additional following condition is satisfied.

$$\text{For each orbit curve } t \to x(t) \text{ of } \mathcal{W}, \text{ the dimension of the tangent vector space } \mathcal{W}(x(t)) \text{ is constant as } t \text{ varies.} \tag{5.3}$$

Then, the submanifolds $C(\mathcal{W}, x_0)$ are maximal integral submanifolds of the singular foliation $\mathcal{W}$ in the sense that the tangent space of $C(\mathcal{W}, x_0)$ at each point $x \in C(\mathcal{W}, x_0)$ is equal to $\mathcal{W}(x)$. Further, (5.3) is satisfied if either X and $\mathcal{W}$ are *real analytic*, or $\mathcal{W}$ is *locally finitely generated* in the sense defined in [16].

There are evident implications for this basic theorem to the study of curvature-zero Poisson structures and the Ehresmann pseudogroups they generate. Let $\{\,,\,\}$ be such an operation on $\mathscr{F}(X)$. For $f \in \mathscr{F}(X)$ let V_f be the vector field on X such that

$$V_f(f') = \{f, f'\} \tag{5.4}$$

for $f \in \mathscr{F}(X)$.

Then, $f \to V_f$ is a Lie algebra homomorphism and

$$\mathscr{G} = \{V_f : f \in \mathscr{F}(X)\}$$

is a Lie subalgebra of $\mathscr{V}(X)$. Hence, Theorems 5.1 and 5.2 apply. $\mathscr{G}$ defines a *singular foliation* of X, i.e., a decomposition into submanifolds. The Lie algebra of vector fields $\mathscr{G}$ (as the pseudogroup it generates) are tangent to the leaves of this foliation.

6. HOMOMORPHISMS OF BIVECTOR FIELDS AND FUNCTION GROUPS IN THE SENSE OF SOPHUS LIE

The Poisson bracket operation of analytical mechanics is usually defined by means of a closed two-differential form, i.e., a twice *covariant* tensor field, a cross-section of the vector bundle $T^d(X) \wedge T^d(X)$. The case where the manifold X is even dimensional and the form is of maximal rank is the traditional one, treated in all the modern treatises on mechanics. The case where the form does not have maximal rank is also interesting, and was first treated partially in [11]. In this case, the Poisson bracket cannot be defined on *all* functions on X, but only on a subalgebra. This "covariant" formalism can also be extended [23] to higher degree differential forms, and thereby to field theories.

Now, covariant tensor fields have certain properties relative to C^∞ mappings between the manifolds on which they live: They "pull-back" dually to the mapping. (This is the meaning of "covariant"!) The Poisson structures arise from contravariant tensor fields, which "push forward". However, the push-forward map cannot be defined for an arbitrary *tensor field*.

To elaborate algebraically, let X and X' be C^∞ paracompact manifolds (possibly of different dimension) and $\phi: X \to X'$ a C^∞ map. At each point $x \in X$, the differential ϕ_* maps the tangent vector space X_x linearly to the tangent space $X'_{\phi(x)}$. This leads to a linear bundle map $\phi_*: T(X) \to T(X')$ and the following commutative mapping diagram:

Lie's Function Groups

$$\phi_*: T(X) \to T(X')$$
$$\downarrow \quad \quad \downarrow$$
$$\phi: X \to X'$$

Consider the cross-sections $\Gamma(T(X))$, $\Gamma(T(X'))$. They do not map naturally under ϕ_*: Given a $V \in \Gamma(T(X))$, $\phi_*(V): X \to \phi_*(V(X))$ cannot be defined naturally as a cross-section of $T(X')$ because a fiber $X'_{x'}$ may arise as the image under ϕ_* of *two* fibers X_{x_1} and X_{x_2}. However, one can impose an extra condition, that $\phi_*(V)$ be well defined as a cross-section of $T(X')$. Let us say that V and V' are ϕ-*related* if:

$$\phi_*(V(x)) = V'(\phi(x)) \tag{6.1}$$

for all $x \in X$,

i.e., if the following diagram of maps is commutative:

$$\begin{array}{ccc} T(X) & \xrightarrow{\phi_*} & T(X') \\ V \uparrow & & \uparrow V' \\ X & \xrightarrow{\phi} & X' \end{array} \tag{6.2}$$

If V and V' satisfy (6.1), then it is readily seen that ϕ *maps orbit curves of* V *into orbit curves of* V', *i.e.*, ϕ *is an intertwining map for the one-parameter pseudogroup of diffeomorphisms generated by* V *and* V'. In this geometric form, the ϕ-related vector fields play a basic role in the Lie-Cartan geometric theory of differential equations. (They are "prolongation maps" for the underlying differential equations.)

Thus, the geometric relation between contravariant tensor fields on X and X' can only be considered as involving special pairs (τ,τ') which are ϕ-related via (6.1)-(6.2). We can do the same for bivector fields.

Definition. Let $\phi: X \to X'$ be a C^∞ map between manifolds, and let $\omega \in \Gamma(T(X) \wedge T(X))$, $\omega' \in \Gamma(T(X') \wedge T(X'))$ be bivector fields on X and X'. They are said to be ϕ-*related*, and we write:

$$\phi_*(\omega) = \omega' \tag{6.3}$$

if the following condition is satisfied:

$$\omega(\phi^*(\theta'_1), \phi^*(\theta'_2)) = \omega'(\theta'_1, \theta'_2) \tag{6.4}$$

for all $x \in X$, all $\theta'_1, \theta'_2 \in X'_{(x)}$.

The following results are easily proved.

Theorem 6.1. Let $\{\,,\,\}_\omega$, $\{\,,\,\}'_{\omega'}$ be the operations on $\mathcal{F}(X)$, $\mathcal{F}(X')$ defined by the bivector fields ω, ω'. Then (6.4) is satisfied, i.e., $\phi_*(\omega) = \omega'$, if and only if the following condition is satisfied:

$$\phi^*(\{f'_1, f'_2\}_{\omega'}) = \{\phi^*(f'_1), \phi^*(f'_2)\}_\omega, \tag{6.5}$$

for all $f'_1, f'_2 \in \mathcal{F}(X')$.

Theorem 6.2. If ω and ω' are ϕ-related, so are their curvature tensors Ω and Ω'. (This says that the differential operator $\omega \to \Omega$ is a "natural" operation.)

7. HOMOMORPHISMS OF POISSON STRUCTURES AND FUNCTION GROUPS IN THE SENSE OF SOPHUS LIE

Let X continue as a manifold with ω a bivector field defining a Poisson operation

$$(f_1, f_2) \to \{f_1, f_2\}$$

on $\mathcal{F}(X)$.

Definition. A set $f_1, \ldots, f_m$ of functions on X is said to form a *function group*, in the sense of Lie, relative to the Poisson tensor, if and only if there are C^∞ functions

$$F_{ij}: R^m \to R, \qquad 1 \leq i, j \leq m,$$

such that:

$$\{f_i, f_j\}_\omega(x) = F_{ij}(f_1(x), \ldots, f_m(x)) \tag{7.1}$$

for all $x \in X$.

Remark. The terminology "function group" is obviously archaic--they are not *groups* in the sense we use the term (except trivially, in the additive structure). However, what Lie usually called a "group" we would call a "Lie algebra". Hence, an appropriate modern name might be *Lie function algebras*.

Theorem 7.1. Suppose that $\phi: X \to (f_1(x), \ldots, f_m(x))$ defines submersion maps from X to R^m. Suppose D is the image in R^m of this submersion; it is, of course, an open subset of R^m. Then, $(f_1, \ldots, f_m)$ form a function group if and only if there is a bivector field ω' on D' such that:

Lie's Function Groups

$$\phi_*(\omega) = \omega' \quad .$$

The proof is given in [16].

8. LIE "FUNCTION GROUPS" GENERATED BY LIE ALGEBRAS OF VECTOR FIELDS

Let us first recall the definition of a "symplectic structure".

Definition. Let X be a manifold. A *symplectic structure* on X is defined by a two-differential form η which satisfies the following conditions:

$$d\eta = 0 \tag{8.1}$$

$$v \lrcorner \eta = 0 \quad \text{for} \quad v \in T(X) \Rightarrow v = 0 \tag{8.2}$$

i.e., η has no nonzero (Cauchy) characteristic vectors.

A symplectic form η defines a Poisson bracket operation on $\mathscr{F}(X)$. For $f \in \mathscr{F}(X)$, define $V_v \in \mathscr{V}(X)$ as follows:

$$df = V_f \lrcorner \eta \quad .$$

Then set:

$$\{f, f_1\} = -V_f(f_1) \quad . \tag{8.3}$$

Theorem 8.1. There is a bivector field ω on X with zero Schouten tensor, which gives rise to the Poisson bracket operation (8.2). Algebraically, is the dual tensor to the (non-singular) η. In local coordinates (x^i), $1 \leq i,j \leq n$, if $\eta = \eta_{ij} dx^i \wedge dx^j$, then

$$\omega = \eta^{ij} \frac{\partial}{\partial x^i} \wedge \frac{\partial}{\partial x^j} \tag{8.4}$$

where (η^{ij}) is the inverse matrix to (η_{ij}) (which exists because of condition (9.2).)

Now, let Q be a manifold and let

$$X = T^d(Q) \tag{8.5}$$

be its cotangent bundle of $\mathscr{V}(Q) \to \mathscr{F}(X)$ defined as follows:

$$f_V(\theta) = \theta(V(q)) \tag{8.6}$$

for $V \in \mathscr{V}(Q)$,

$\theta \in Q_q$, $q \in Q$.

Let η be the canonical symplectic form on $T^d(Q) \equiv X$, and let $\{\,,\,\}$ be the corresponding Poisson bracket on $\mathscr{F}(X)$. (It is, of course, the standard Poisson bracket used in analytical mechanics when Q is the configuration space manifold of the mechanical system.)

Theorem 8.2. For $V_1, V_2 \in \mathscr{V}(Q)$, then

$$\{f_{V_1}, f_{V_2}\} = f_{[V_1, V_2]} \tag{8.7}$$

i.e., the mapping $V \to f_V$ is a Lie algebra homomorphism from $\mathscr{V}(Q)$ to the Lie algebra (and Poisson bracket) $\mathscr{F}(T^d(Q))$.

Proof. Well known.

Suppose now that $\mathscr{G}$ is a finite dimensional Lie algebra of vector fields on Q. Let

(V^a), $1 \leq a, b, c \leq m$

be a basis for $\mathscr{G}$. It satisfies relations of the following form:

$$[V^a, V^b] = \lambda_c^{ab} V^c, \tag{8.8}$$

where (λ_c^{ab}) are the *structure constants* of the Lie algebra.

By Theorem 8.2, the functions $f^a \equiv f_{V^a}$ on $T^d(Q) = X$ satisfy the same Poisson-bracket relations:

$$\{f^a, f^b\} = \lambda_c^{ab} f^c. \tag{8.9}$$

Theorem 8.3. Let D be an open subset of X on which the differentials

$df^1, \ldots, df^m$

have constant rank. Let $\phi: D \to R^m$ be the following map:

$x \to (f^1(x), \ldots, f^m(x))$.

Then, there is a Poisson structure on the submanifold $\phi(D)$ of R^m. If the indices are relabelled so that the $df^1,\ldots,df^m$ are a maximal linearly independent set among the (df^a) on D, then the $(f^1,\ldots,f^n)$ form a function group in the sense of Lie.

<u>Proof</u>. The $f^{m+1},\ldots,f^m$ can be written locally as functions of the $f^1,\ldots,f^n$.

References

1. R. Hermann, The geometric foundations of the integrability property of differential equations and physical systems, I: Lie's "function groups", (to appear, *J. Math. Phys.*).

2. S. Lie, *Transformationsgruppen*, Vol. II, Chelsea Publishing Co., New York.

3. R. Hermann, The Lax representation as a "quantization" of the function groups of Sophus Lie, *Phys. Rev. D* <u>26</u>, 1491-1492 (1982).

4. A. Lichnerowicz, Les varieties de Poisson et leurs algebras de Lie associees, *J. Diff. Geom.* <u>12</u>, 253-300 (1977).

5. J. Plante, Foliations with measure preserving holonomy, *Ann. of Math.* <u>102</u>, 327-361 (1975).

6. R. Hermann, Current algebras, the Sugawara model, and differential geometry, *J. Math. Phys.* <u>11</u>, 1825-1829 (1970); Geometric formula for current-algebra commutation relations, *Phys. Rev.* <u>177</u>, 2449 (1969); Quantum field theories with degenerate Lagrangians, *Phys. Rev.* <u>177</u>, 2453 (1969). *AIP Conference Proceedings No. 88*, p. 67, 1982.

7. J.A. Schouten, On the differential operators of first order in tensor calculus, *Convengo Intern. Geometria Differenziale Italia*, 1953, Ed. Cremonese, Roma, 1954, 1-7.

8. A. Nijenhuis, Jacobi-type identities for bilinear differential concomitants of certain tensor fields, *I. Indag. Math.* <u>17</u>, 390-403 (1955).

9. R. Hermann, *Interdisciplinary Mathematics*, Vol. 10, Chaps. 14 and 15, Math Sci Press, Brookline, MA, 1975.

10. R. Hermann, *Interdisciplinary Mathematics*, Vol. 14, Math Sci Press, Brookline, MA, 1977.

11. R. Hermann, *Geometry, Physics and Systems*, Marcel Dekker, New York, 1973.

12. C. Ehresmann, Sur les structures infinitesimals regulieres, *Congres Intern. Math. Amsterdam*, Vol. 1, 479-480 (1954); Connexions infinitesimales, *Colloque Top. Alg. Bruxelles*, 29-55 (1950); Structures infinitesimales et pseudogroupes de Lie, *Colloq. Intern. C.N.R.S. Geom. Diff. Strasbourg*, 97-110 (1953); *Compte-rendus Acad. Sc. Paris* <u>240</u>, 1762 (1954); <u>241</u>, 397 and 1955 (1955); <u>246</u>, 360 (1958); Connexions d'ordre superieur, *Atti 5 Congr. dell'Unione Mat. Italiana 1955*, Ed. Cremonese, Roma, 326-328 (1956); Categories topologiques et categories differentiables, *Colloq. Geom. Diff. Globale Bruxelles*, C.B.R.M., 137-150 (1958); Groupoides differenciales, *Rivista Un. Mat. Argentina* XIX, Guenos-Aires, 48 (1960).

13. C. Caratheodory, Untersuchungen uber die Grundlagen der Thermodynamik, *Math. Ann.* <u>67</u>, 355-386 (1909).

14. W.L. Chow, Uber Systeme von Linearen Partiellen Differentialgleichungen Erster Ordnung, *Math. Ann.* 117, 98-105 (19).

15. R. Hermann, On the accessibility problem in control, *Internat. Sympos. Nonlinear Differential Equations and Nonlinear Mechanics*, Academic Press, New York, 1963, 325-332.

16. R. Hermann, Cartan connections and the equivalence problems for geometric structures, *Contributions to Differential Equations* 3, 199-248 (1964).

17. P. Stefan, Two proofs of Chow's theorem, in *Geometric Methods in System Theory*, D. Mayna and R. Brockett (eds.), D. Reidel, 1953.

18. H.J. Sussmann, Orbits of families of vector fields and integrability of systems with singularities, *Bull. Amer. Math. Soc.* 79, 197-199 (1973).

Chapter 7

PICARD-VESSIOT THEORY

1. INTRODUCTION

The Picard-Vessiot theory of linear ordinary differential operators is one of the glories of 19th century mathematics. It had both an algebraic and geometric side; the former had roots in the Galois theory of algebraic equations, the latter in Lie's theory of transformation groups. The algebraic side has been developed in the 20th century, while the geometric aspects have not been extensively pursued. In this work I will sketch some ideas which will be developed more fully later on.

In addition to its general mathematical interest (e.g., as a theory which decides in principle whether differential equations can be solved in terms of certain classes of functions and operations in those functions), it might play a role in the theory of nonlinear waves--isospectral deformation and in the theory of isomonodromy deformation.

2. DIFFERENTIAL EQUATIONS IN FIBER SPACES, AND JET SHEAVES OF SOLUTIONS

Let us now present (in simplified form) some concepts from the modern geometric theory of differential equations. Let X and E be finite dimensional, real-analytic, paracompact manifolds. Let

$$\pi: E \to X$$

be a submersion map of E *onto* X. We will call the triple

$$(E, \pi, X)$$

a *fiber space*.

If U is an open subset of X, let

$$\Gamma_U(\pi) \tag{2.1}$$

be the space of real-analytic cross-section maps

$$\gamma: U \to E \tag{2.2}$$

Consider the set of ordered triples

$$(x, U, \gamma) \tag{2.3}$$

where:

$$x \in X, \tag{2.4}$$

$$U \text{ is an open neighborhood of } x \text{ in } X, \tag{2.5}$$

$$\phi \in \Gamma_U . \tag{2.6}$$

Introduce an equivalence relation $\sim$ into the set defined by relations (2.3)-(2.6) as follows:

$$(x, U, \gamma) \sim (x', U', \gamma')$$

iff

$$\begin{aligned} x &= x' \\ \gamma &= \gamma' \end{aligned} \tag{2.7}$$

$$U \cap U' \tag{2.8}$$

The quotient of the set of triples (2.3) is denoted by:

$$\underline{\Gamma}(\pi)$$

and is called the *sheaf of germs of cross sections of maps of the fiber space* π. The projection

$$(x, U, \gamma) \to x$$

passes to the quotient (because of (2.6) and defines a mapping that we denote by

$$\underline{\pi} : \underline{\Gamma}(\pi) \to X .$$

The usual sheaf-theoretic topology is put on $\underline{\Gamma}(\pi)$, i.e., the topology generated by considering the image of Γ_U as the open subsets where U runs through the set of open subsets of X. Continuous cross-sections of $\underline{\pi}$ can then be identified with $\Gamma(\pi)$ itself.

For each integer $k \geq 0$, introduce another equivalence relation into the space of all triples

$$(x, U, \gamma) .$$

Namely,

$$(x, U, \gamma) \sim (x', U', \gamma') \tag{2.9}$$

if the following conditions are satisfied:

$$x = x' \tag{2.10}$$

γ and γ' agree to the k-th order at x, i.e., the first k-th order derivation in any local coordinate system for X agree at x.

The quotient by this equivalence relation is denoted by:

$$J^k(\pi) \tag{2.11}$$

i.e., the space of k-jets of cross-sections of π. Then, the $J^k(\pi)$ are real analytic manifolds, and there is a tower of submersion maps:

$$X \leftarrow E = J^0(\pi) \leftarrow J^1(\pi) \leftarrow \cdots \tag{2.12}$$

3. DIFFERENTIAL EQUATIONS

Keep the notation of Section 2. $\pi: E \to X$ is a fiber space map. A k-th order *differential equation* is an analytic subset D of $J^k(\pi)$. A cross-section $\gamma \in \Gamma_U(\pi)$ is a *solution of* D

$$j^k(\gamma) \subset D . \tag{3.1}$$

The germs of solution from a subset of the sheaf $\underline{\Gamma}(\pi)$, denoted as

$$\underline{D} ,$$

which, for many geometric and physical purposes can be identified with D itself. (Of course, in category language $D \to \underline{D}$ is a functor.)

4. PROLONGATION

Continue with the setting described in Sections 2 and 3. Let

$$\pi: E \to X$$

$$\pi': E \to X$$

be two submersion maps with the same base space X. Let $D \subset J^k(E)$, $D' \subset J^{k'}(E')$ be differential equations of order k, k', respectively.

Let

$$\phi: E \to E'$$

be a fiber-preserving map of E onto E', i.e., there is a map $\phi_X: X \to X$ with a commutative diagram

$$\begin{array}{ccc} E & \xrightarrow{\phi} & E' \\ \pi \downarrow & & \downarrow \pi' \\ X & \xrightarrow{\phi_X} & X' \end{array}$$

ϕ maps cross-sections of π into cross sections of π':

If $\gamma \in \Gamma(\pi)$, then

$$\phi(\gamma): x \to \phi(\gamma)\phi_X^{-1}(x)$$

is a cross-section of π'.

This operation on cross-sections passes to the quotient to define a map

$$\underline{\phi}: \underline{\Gamma}(\pi) \to \underline{\Gamma}(\pi')$$

of the sheaf of cross-sections of π to the sheaf of cross-sections to π'.

Definition. The fiber preserving map $\phi: E \to E'$ defines the differential equation D as a *prolongation* of the differential equation D' if the induced mapping $\underline{\phi}$ on the sheaf of solutions $\underline{D}$ is *onto* $\underline{D'}$.

This definition (in more-or-less the form given above) plays a basic role in the work of Lie, Vessiot and Cartan. Of course, a special case is that where

$$E = E',$$
$$\pi = \pi'.$$

is then a *fiber space automorphism*. The set of all fiber space automorphisms form a group, called the group of *gauge transformations*.

If G is a group of fiber space automorphisms, it acts on the triples

$$(x, U, \gamma)$$

of form (2.10), and preserves the equivalence relation (2.10), hence acts on

Picard-Vessiot Theory 101

the jet fiber spaces $J^k(\pi)$. These actions are called *prolongations* of the action of G.

5. GROUPS OF SYMMETRIES OF DIFFERENTIAL EQUATIONS AND SYMMETRIES IN THE PROLONGATION SENSE

Let D be a differential equation defined on a fiber space (E,π,X). Let G be a group of fiber space automorphisms. G acts also on the sheaf of cross-sections.

<u>Definition</u>. G is a group of symmetries of the differential equation D if, when made to act on the sheaf $\Gamma(\pi)$, G maps solutions of D onto itself.

Now, suppose that (E',π',X) is another fiber space with a differential equation D' defined for E. Let G be a group of symmetries for D', and let $\phi: E \to E'$ be a map that defines D' as a prolongation of D.

<u>Definition</u>. G is said to act on $\underline{D}$ as a group of *prolongation symmetries*.

6. DIFFERENTIAL OPERATORS AND DIFFERENTIAL INVARIANTS AND COVARIANTS

Let

$$(E,X,\pi)$$

be a fiber space, as before. Let $J^k(\pi)$ be the k-jets of cross-sections, considered as a fiber space over X. Let (E',X,π') be another fiber space with X as base. Let G be a group which acts on both E and E' via fiber preserving automorphisms. G also acts on $J^k(\pi)$, via a prolonged action. Let

$$\Delta: J^k(\pi) \to E' \qquad (6.1)$$

be a fiber preserving map. It is called a *differential operator* associated with the fiber space π.

<u>Definition</u>. Δ is a *differential covariant* of the action of G if, as a map with domain and range indicated in (6.1), it intertwines the action of G. In the special case that G acts as the identity on E', Δ is called a *differential invariant* of the action of G.

7. A GENERAL FRAMEWORK FOR PICARD-VESSIOT

Let $\pi: E \to X$ continue as a fixed fiber space. Let $\pi': E' \to X$ be another fiber space, and let

$$\phi: E' \to E$$

be a fiber preserving map that defines π' as a prolongation of π. Let D and D' be differential equations for cross-sections of E and E', and suppose ϕ defines D' as a prolongation of D. In addition, suppose the following data is given:

1) A group G of automorphisms of D'

2) A collection of differential operators

$$\{\Delta\}: J^k(E) \to E''$$

($\pi''\colon E'' \to X$ is a third fiber space)

Suppose that, for each $\Delta \in \{\Delta\}$, each solution $\gamma \in \underline{D}$, the differential equation, the map

$$\Delta j^k(\gamma)$$

of $X \to E''$ belongs to a *given* class of mappints that we call *rational*.

<u>Definition</u>. The subgroup of the $g \in G$ whose prolongation to $J^k(E)$ leaves invariant each operator $\Delta \in \{\Delta\}$ is called the *Picard-Vessiot group*.

Let us turn to the traditional example.

8. THE GEOMETRIC SETTING FOR THE CLASSICAL PICARD-VESSIOT THEORY

Let X be an open subset of the complex numbers $\mathbb{C}$. Let

$$E = X \times \mathbb{C}$$
$$= \pi(x,y)$$
$$= x,$$
$$x \in X, \quad y \in \mathbb{C}.$$

For each open subset U of X, $\Gamma_U(E)$ is then the space of real analytic maps $y\colon U \to V$. Let D be the differential equation defined by the Cauchy-Riemann equation, plus the linear ordinary differential equation.

$$a_n(x) \frac{d^n y}{dx^n} + \cdots + a_0(x) y = 0 \qquad (8.1)$$

where:

$$x \to a_n(x), \ldots, a_0(x)$$

Picard-Vessiot Theory 103

are complex analytic maps of $X \to \mathbb{C}$. Now, let

$$E' = X \times GL(n,\mathbb{C})$$
$$= \{(x,M)\} \ .$$

($GL(n,\mathbb{C})$ = group of invertible $n \times n$ complex matrices.) Let D' be the linear matrix differential equation:

$$\frac{dM}{dx} = A(x)M \qquad (8.2)$$

where

$$A = \begin{pmatrix} a_{11}(x), \ldots, a_{1n}(x) \\ \vdots \\ a_{n1}(x), \ldots, a_{nn}(x) \end{pmatrix}$$

plus the Cauchy-Riemann equations. Let $\phi: E' \to E$ be of the form:

$$\phi(x,M) = \alpha_1 M \alpha_2 \qquad (8.3)$$

α_1 is an $1 \times n$, α_2 and $n \times 1$ matrix.

The "prolongation" condition amounts to saying that, for each solution

$$x \to M(x)$$

of (8.2), the function

$$y(x) = \alpha_1 M(x) \alpha_2 \qquad (8.4)$$

is a solution of (8.1)

Then (using subscripts for derivatives),

$$y_x = \alpha_1 M_x \alpha_2$$
$$= \text{, using (8.2),}$$
$$\alpha_1 A M \alpha_2 \qquad (8.5)$$

$$y_{xx} = \alpha_1 (A_x M + A M_x) \alpha_2$$
$$= \alpha_1 (A_x M + A^2 M) \alpha_2$$
$$= \alpha_1 (\nabla(A) M) \alpha_2 \qquad (8.6)$$

where

$$\nabla A = A_x + A^2 \,. \tag{8.7}$$

($A \to \nabla A$ is a sort of "covariant derivative" operation.)

$$\begin{aligned} y_{xxx} &= \alpha_1 (\nabla(A)_x M + (\nabla A) M_x) \alpha_2 \\ &= \alpha_1 ((\nabla A)_x M + (\nabla A) A M) \alpha_2 \\ &= \alpha_1 \nabla^2 A M \alpha_2 \end{aligned} \tag{8.8}$$

where

$$\nabla^2 A = (\nabla A)_x + (\nabla A) A \tag{8.9}$$

Continuing in this way, we have, by induction on n:

$$\frac{d^n y}{dx^n} = \alpha_1 \nabla^{n-1}(A) M \alpha_2 \tag{8.10}$$

where

$$\nabla^{n-1}(A) = (\nabla^{n-2} A)_x + (\nabla^{n-2} A) A \tag{8.11}$$

Then,

$$a_n \frac{d^n y}{dx^n} + \cdots + a_0 y = a_n \alpha_1 \nabla^{n-1}(A) M \alpha_2 + \cdots \tag{8.12}$$

The condition that y satisfy the differential equation (8.1) is then that:

$$\alpha_1 (a_n \nabla^{n-1}(A) + a_{n-1} \nabla^{n-2}(A) + \cdots a_0) M \alpha_2 = 0 \tag{8.13}$$

Since M is arbitrary, this implies that:

$$\alpha_1 (a_n \nabla^{n-1}(A) + a_{n-1} \nabla^{n-2}(A) + \cdots + a_0) = 0 \tag{8.14}$$

Then, there is finally the prolongation condition.
 For example, let us consider the case where:

$$n = 2 \tag{8.15}$$

Then, (8.14) takes the form:

Picard-Vessiot Theory

$$\alpha_1(a_2(A_x + A^2) + a_1 A + a_0) = 0 \ . \tag{8.14}$$

One way of satisfying this relation is to assume that the following pair of equations is satisfied:

$$a_2 A^2 + a_1 A + a_0 = 0 \tag{8.15}$$

$$\alpha_1 A_x = 0 \ . \tag{8.16}$$

(8.15) implies that a_0, a_1, a_2 are determined by the characteristic polynomial of A, while (8.16) restricts A_x to have a *zero eigenvalue*.

Of course, we can also find such a prolongation by following the classical procedure for converting the n-th order scalar equation (8.1) into first order matrix equation. Let us do this for $n = 2$. The equation (8.1) is then:

$$a_2 y_{xx} + a_1 y_x + a_0 y = 0 \ . \tag{8.17}$$

Set:

$$y_1 = y$$

$$y_2 = y_x \tag{8.18}$$

$$y_{1,x} = y_2$$

$$y_{2,x} = -a_1 a_2^{-1} y_2 - a_0 a_2^{-1} y_1 \ ,$$

or

$$\begin{pmatrix} y_1 \\ y_2 \end{pmatrix}_x = \begin{pmatrix} 0 & , & 1 \\ -a_0 a_2^{-1} & , & -a_1 a_2^{-1} \end{pmatrix} \begin{pmatrix} y_1 \\ y_2 \end{pmatrix} \tag{8.19}$$

$$= A \begin{pmatrix} y_1 \\ y_2 \end{pmatrix} \tag{8.20}$$

with

$$A = \begin{pmatrix} 0 & , & 1 \\ -a_0 a_2^{-1} & , & -a_1 a_2^{-1} \end{pmatrix} \tag{8.21}$$

This can now be converted into a matrix system:

$$M_x = AM \tag{8.22}$$

$$M = \begin{pmatrix} y_1 & y_1' \\ y_2 & y_2' \end{pmatrix} \tag{8.23}$$

if:

$$\alpha_1 = (1,0) \tag{8.24}$$

$$\alpha_2 = \begin{pmatrix} 1 \\ 0 \end{pmatrix} \tag{8.25}$$

then

$$\alpha_1 M \alpha_2 = (1,0) \begin{pmatrix} y_1 \\ y_2 \end{pmatrix}$$

$$= y_1$$

$$= \quad , \text{ using } (8.18),$$

$$y \quad .$$

Let us sum up as follows.

<u>Theorem 8.1.</u> If the 2×2 matrix M is a solution of the first order matrix differential equation (8.22), then

$$y = \alpha_1 M \alpha_2$$

is a solution of the second order scalar equation (8.17). if

$$E' = X \times GL(2,\mathbb{C})$$

$$E' = X \times \mathbb{C} \quad ,$$

then the map

$$\alpha: (x,M) \to (x, \alpha_1 M \alpha_2)$$

defines a prolongation of differential equations.

Note that:

$$A_x = \begin{pmatrix} 0 & , & 0 \\ -a_0 a_2^{-1} & , & -a_1 a_2^{-1} \end{pmatrix}_x$$

so that A_x is nilpotent, i.e., has zero eigenvalues.

9. THE CLASSICAL PICARD-VESSIOT THEORY

Return to the case of an n-ordinary, linear matrix equation:

$$M_x = AM \ . \tag{9.1}$$

Let E be the product bundle of base X (an open subset of the complex plane) and fiber $GL(n,\mathbb{C})$, i.e., a point of E is a pair

$$(x,M) \tag{9.2}$$

$$x \in \mathbb{C}, \quad m \in GL(n,\mathbb{C})$$

$$\pi(x,M) = x \ . \tag{9.3}$$

(9.1) then defines a differential equation for cross-sections of E.

$GL(n,\mathbb{C})$ acts as a group of "gauge transformations" on E, i.e., a group of fiber space automorphisms which acts on each fiber:

$$(x,M) \to (x, Mg^{-1}) \tag{9.4}$$

$$x \in X ; \quad M,g \in GL(n,\mathbb{C}) \ .$$

This action groups a solution of (9.1) onto another solution, i.e., acts as a group of symmetries. This action is *transitive* and *free*, i.e., the space of solutions is identified with $GL(n,\mathbb{C})$ itself. (This is then what Vessiot defined as a *Lie system*.)

The k-th (holomorphic) jet bundle of E, $J^k(E)$, can be identified with the space of (k+2)-triples:

$$(x, M, M^{(1)}, \ldots, M^{(k)}) \tag{9.5}$$

$$x \in X, \quad M \in GL(n,\mathbb{C}), \quad M^1, \ldots, M^{(k)} \in L(\mathbb{C}^n, \mathbb{C}^n) \ .$$

($L(\mathbb{C}^n, \mathbb{C}^n)$ is the space of $n \times n$ complex matrix matrices identified with the vector of space of linear maps: $\mathbb{C}^n \to \mathbb{C}^n$. It is the Lie algebra of $GL(n,\mathbb{C})$ in the well known way.) If

$$\gamma: x \to (x, M(x))$$

is a (holomorphic) cross-section of E, its k-jet is

$$j^k(\gamma): \gamma \to (x, M(x), M_x, M_{xx}, \ldots) \qquad (9.6)$$

Let us now consider differential operators

$$\Delta: J^k(E) \to \mathbb{C}$$

which do not depend on x, i.e., maps:

$$M, M^1, \ldots, M^{(k)} \to \Delta(M, M^1, \ldots, M^k) \qquad (9.7)$$

$$GL(n, \mathbb{C}) \times L(\mathbb{C}^n, \mathbb{C}^n) \times \cdots \times L(\mathbb{C}^n, \mathbb{C}^n) \to \mathbb{C}.$$

Now, fix in advance a certain class of mappings of $X \to X$ that we call *rational*. (They could, for example, be rational maps in the usual sense.) Consider the class of Δ's which satisfy the following condition:

$$\Delta j^k(\gamma): X \to X$$

is a rational map in the sense prescribed, for *every solution* $\gamma: X \to (x, M(x))$ of (9.1). Let us call the collection of such Δ's (varying the integer k also) the *Picard-Vessiot operators*.

Definition. The *Picard-Vessiot group* of the differential equation (9.1) related to the fixed rational maps $X \to \mathbb{C}$) is the set of all elements of the group $GL(n, \mathbb{C})$ (acting as above on the solutions of (9.1)) which leaves *invariant* all Picard-Vessiot operators.

What is classically the most interesting feature of the Picard-Vessiot group in this case is its relation to the classical *monodromy group*. In fact, the Picard-Vessiot group typically contains the monodromy group as a discrete subgroup, and, in favorable cases (e.g., regular singularities) is its *algebraic closure*.

References

1. L. Schlesinger, *Handbuch der Theories der Lineardifferentialgleichungen*, B.G. Teubner, Leipzig, 1897.

2. A.P. Fordy and J. Gibbons, *J. Math. Phys.* **21**, 2508 (1980).

3. E. Picard, *Traité d'Analyse*, Gauthier-Villars, Paris, 1928 (Vol. 3, Chapter 7).

4. E. Vessiot, Sur les integrations des equations differentielles lineaires, *Am. Sci. Ecole Norm.* 131(9), 192-280 (1892).

5. F. Kolchin, *Differential Algebra and Algebraic Groups*, Academic, New York, 1973.

6. I. Kaplansky, *Introduction to Differential Algebra*, Herman, Paris, 1957.

Chapter 8

RELATIONS BETWEEN KORTEWEG-DE VRIES AND PICARD-VESSIOT THEORY

Abstract

As is well known, nonlinear wave theory is closely linked to the algebra of differential operators in one variable. The classical Picard-Vessiot theory is basically a theory of factorization of differential operators. Certain general relations will be described, leading to a natural description of the Muira transformation and the associated Backlund transformation.

1. INTRODUCTION

Through the work of Lax, Gelfand, Dikii, Krichever, and others, the relations between the algebra of differential and pseudo-differential operators in one variable and the theory of nonlinear waves has been extensively developed. What has been used in this work is the *additive* and *Lie algebra* structure of the algebra of differential operators.

The classical Picard-Vessiot theory also has much to say about the algebra of differential operators--it emphasizes the *multiplicative* structure of the algebra. On general grounds, one might think that it too should play a role in the theory of nonlinear waves. The purpose of this paper is to describe one such relation. We show that the Muira transformation and the associated Backlund transformation of the Korteweg-de Vries are most naturally interpreted in terms of the Picard-Vessiot theory.

2. AN ALGEBRAIC SETTING. LAX STRUCTURES

Let us begin with an algebraic model for the algebra of differential operators in one variable. Let $\mathcal{D}$ be an associative algebra unit element. (The scalar field will be the real or complex numbers.)

A subset $\mathcal{S}$ of a vector space is said to be affine if the set of differences

$$\{s_1 - s_2 : s_1, s_2 \in \mathcal{S}\}$$

of elements of $\mathcal{S}$ is a linear subspace of the vector space.

Consider an affine subspace $\mathcal{D}'$ of $\mathcal{D}$. Let L be the linear subspace formed by the difference between elements of $\mathcal{D}'$. Consider a map (not necessarily linear) $\alpha: \mathcal{D}' \to \mathcal{D}$ such that:

$$[D, \alpha(D)] \in L \qquad (2.2)$$

for $D \in \mathcal{D}'$

One can then form a vector field $V: \mathcal{D}' \to \mathcal{D}$ as follows:

$$V(D) = [D, A(D)] \qquad (2.3)$$

for $D \in \mathcal{D}'$.

One can then form the orbits of the vector field V, i.e., the solution of the following enduction equation:

$$\frac{dD}{dt} = V(D(t))$$
$$= [D(t), \alpha(D(t))] \qquad (2.4)$$
$$D(t) \in D.$$

This is called the *Lax equation* determined by the map α. (All analysis will be "formal", i.e., it will be assumed that differential equations have solutions, limits can be freely interchanged, etc. The purpose of this section is to provide algebraic motivation for geometric and analytic phenomena.)

Let G be the group of automorphisms of the algebra $\mathcal{D}$. We see that the solutions of (2.4) are orbits of curves in G, i.e., there is a curve $t \to g(t)$ in G such that the solution of (2.4) is of the form

$$D(t) = g(t)(D(0)) \qquad (2.5)$$

(Of course, this is also "formal".)

<u>Definition</u>. A *Lax structure* (within the algebra $\mathcal{D}$) is a triple: $(\mathcal{D}', \alpha, L)$ such that:

a) $\mathcal{D}'$ is an affine subspace of $\mathcal{D}$,

b) L is a linear subspace of $\mathcal{D}$ such that

$$L = \{D_1 - D_2 : D_1, D_2 \in \mathcal{D}'\}$$

c) α is a map $\mathcal{D}' \to \mathcal{D}$ such that

$$[D, \alpha(D)] \in L$$

for all $D \in \mathcal{D}'$

Korteweg-De Vries   113

3. SOME EXAMPLES OF LAX SYSTEMS

Continue with $\mathcal{D}$ as an associative algebra with a unit.

Let δ be an element of $\mathcal{D}$, and let L be a linear subspace of such that the following conditions are satisfied:

$$[\delta, L] \subset L \qquad (3.1)$$

For a L, write:

$$\delta(a) = [\delta, a] \qquad (3.2)$$

$$LL \subset L , \qquad (3.3)$$

i.e., L is a subalgebra of $\mathcal{D}$.

$$[a,b] = 0 \qquad (3.4)$$

for $a, b \in L$,

i.e., L is an abelian subalgebra of $\mathcal{A}$.

For each $a \in \mathcal{A}$, there is an $f \in \mathcal{A}$ such that:

$$\delta(f) = af \qquad (3.5)$$

f^{-1} also exists.

If $\delta(a) = 0$ for $a \in L$, then a is a scalar multiple of the unit of the algebra $\mathcal{D}$.

Form an affine subspace of $\mathcal{D}$ as follows:

$$\mathcal{D}' = \{\delta + a : a \in L\} . \qquad (3.6)$$

For $D = \delta + a \in \mathcal{D}'$, we can write:

$$D = f^{-1} \delta f \qquad (3.7)$$

where f satisfies (3.4).

In order to define a Lax system, we must find a map $\alpha: \mathcal{D}' \to \mathcal{D}$ such that

$$[D, \alpha(D)] \in L \qquad (3.8)$$

for all $D \in \mathcal{A}$.

Let us look for $\alpha(D)$ of the following form:

$$\alpha(D) = \delta^n + b_{n-1}\delta^{n-1} + \cdots + b_0 \tag{3.9}$$

with $b_0, \ldots, b_{n-1} \in L$.

Using (3.7), we have:

$$[f^{-1}\delta f, \alpha(D)] \in L \tag{3.10}$$

We can rewrite (3.10) as follows:

$$[\delta, f\alpha(D)f^{-1}] \in L$$

Now, $f\alpha(D)f^{-1}$ is of form similar to (3.9), i.e.,

$$f\alpha(D)f^{-1} = \delta^n + c_{n-1}\delta^{n-1} + \cdots + c_0 . \tag{3.11}$$

Thus, we have

$$[\delta, f\alpha(D)f^{-1}] = \delta(c_{n-1})\delta^{n+1} + \cdots + \delta(c_0) .$$

This suggests that we choose $\alpha(D)$ so that the following condition is satisfied:

$$\delta(c_{n-1}) = 0 = \cdots = \delta(c_1) ,$$

i.e., the coefficients above the zeroth of $f\alpha(D)f^{-1}$ are scalar. We can then formalize our finding as follows.

Theorem 3.1. The following formula defines a Lax system: for $D = \delta + a$: $a \in L$

$$\alpha(D) = f^{-1}(c_{n-1}(a)\delta^n + \cdots + c_1(a)\delta)f + \beta_0(a) \tag{3.12}$$

where $c_1, \ldots, c_{n-1}$ are mappings of L to the scalar field, and where β_0 is a mapping: $L \to L$. f is any $f \in L$ such that $f^{-1}\delta(f) = a$. The right hand side of (3.12) is independent of the choice of f.

We can then write down the Lax equation for this Lax system.

$$\frac{da}{dt} = [f^{-1}\delta f, f^{-1}(c_{n-1}(a)\delta^n + \cdots + c_1(a)\delta)f^{-1} + \beta(a)]$$
$$= \beta(a) . \tag{3.13}$$

In order to have more explicit formulas, let us exploit Leibniz's formula:

Korteweg-De Vries

$$\delta^n f = \sum_{j=0}^{n} \binom{n}{j} \delta^{n-j}(f) \delta^j \qquad (3.14)$$

This enables us to compute $\alpha(D)$ in the form

$$\alpha(D) = \delta^n + \alpha_{n-1}(a)\delta^{n-1} + \cdots + \alpha_0(a) \qquad (3.15)$$

where $\alpha_{n-1}, \ldots, \alpha_0$ are maps: $L \to L$.

Remark. That one could build up Lax systems based on first order differential operators was mentioned (based on a different method) in IM 14, p. 142.

4. THE SKEW-ADJOINTNESS CONDITION FOR FIRST ORDER LAX SYSTEMS

Suppose now that the algebra $\mathcal{D}$ has a linear map

$$D \to D^*$$

which satisfies the following conditions:

$$(D_1 D_2)^* = D_2^* D_1^* \qquad (4.1)$$

for $D_1, D_2 \in \mathcal{D}$.

$$(D^*)^* = D \qquad (4.2)$$

for $D \in \mathcal{D}$.

(As the notation suggests, an example of such an operator is the "Lagrange adjoint" for differential operators.) Suppose that the data used to define a Lax structure satisfies the following condition:

$$\delta^* = -\delta \qquad (4.3)$$

$$a^* = a$$

for $a \in \mathcal{L} \qquad (4.4)$

As in Section 3, define a Lax system:

$$\mathcal{D}' = \{\delta + a : a \in L\} \qquad (4.5)$$

$$\alpha(\delta + a) = \delta^n + \alpha_{n-1}(a)\delta^{n-1} + \cdots + \alpha_0(a) \qquad (4.6)$$

where $\alpha_{n-1}, \ldots, \alpha_0$ are maps: $L \to L$. Require that $(\mathcal{D}', L, \alpha)$ define a Lax system. In addition, suppose that the Lax system be *skew-adjoint*, in the sense that:

$$\alpha(\delta + a)^* = -\alpha(-\delta + a) \tag{4.7}$$

$$= \alpha(\delta - a)$$

for all $a \in L$.

Now, by Theorem 3.1, α is of the following form:

$$\alpha(\delta + a) = f^{-1}(\delta^n + c_{n-1}(a)\delta^{n-1} + \cdots + c_1(a)\delta)f + \beta(a) \tag{4.8}$$

where $c_1, \ldots, c_{n-1}$ are scalars.

First of all, (4.7) requires that:

$$n \text{ is odd} \tag{4.9}$$

Thus,

$$\alpha(\delta + a)^* = f(-\delta^n + c_{n-1}(a)\delta^{n-1} - \cdots - c_1(a)\delta)f^{-1} + \beta(a)$$

$$= \alpha(\delta - a)$$

$$= f^{-1}(\delta^n + c_{n-1}(-a)\delta^{n-1} + \cdots +) + \beta(-a)$$

Hence,

$$f^{-2}(\delta^n + c_{n-1}(-a)\delta^{n-1} + \cdots + c_1(-a)\delta)f^2$$

$$+ (-\delta^n + c_{n-1}(a)\delta^{n-1} - \cdots - c_1(a)\delta)$$

$$+ \beta(a) - \beta(-a)$$

$$= 0 \tag{4.10}$$

This is the basic relation determining the Lax system.

Explicit Calculation for $n = 3$

It is just as easy to work directly, rather than use the general formula

$$\alpha(\delta + a) = \delta^3 + \alpha_2 \delta^2 + \alpha_1 \delta + \alpha_0$$

$$\alpha(\delta + a)^* = -\delta^3 + \delta^2 \alpha_2 - \delta\alpha_1 + \alpha_0$$

$$= -\delta^3 + \delta^2(\alpha_2) + 2\delta(\alpha_2)\delta + \alpha_2\delta^2 - \alpha_1\delta - \delta(\alpha_1)$$

$$= -\delta^3 - \alpha_2\delta^2 - \alpha_1\delta - \alpha_0$$

Korteweg-De Vries 117

or

$$\alpha_2 = 0$$
$$\alpha_0 = \delta(\alpha_1) \ .$$

Hence, $\alpha(\delta + a)$ is of the following form:

$$\alpha(\delta + a) = \delta^3 + \alpha_1 \delta + \delta(\alpha_1) \ .$$

Thus,

$$[\delta + a, \alpha(\delta + a)] = \delta(\alpha_1)\delta + \delta^2(\alpha_1) + [a, \delta^3] + \alpha_1[a, \delta]$$

$$= \delta(\alpha_1)\delta + \delta^2(\alpha_1) - (\delta^3(a) + 3\delta^2(a)\delta + 3\delta(a)\delta^2)$$

$$- \delta(a)\alpha_1 \ .$$

The Lax condition requires that all terms of order ≥ 1 on δ vanish.

5. THE MUIRA TRANSFORM

Now we shall present a construction we call the "Muira transform", since it in some sense generalizes the now classic transform discussed by Muira between the Korteweg-de Vries and modified Korteweg-de Vries equation.

Consider an associated algebra $\mathcal{D}$, with a unit element and an adjoint operation $D \to D^*$. Suppose given one Lax structure $(\mathcal{D}', L,)$. Recall that this means that:

α is a map $\mathcal{D}' \to \mathcal{D}$ such that

$$[D, \alpha(D)] \in L \qquad (5.1)$$

for $D \in \mathcal{D}'$.

One can construct an *adjoint Lax system* as follows. It is denoted by

$(\mathcal{D}'^*, L^*, \alpha^*)$

$$\mathcal{D}'^* = \{D^* : D \in \mathcal{D}\} \qquad (5.2)$$

$$L^* = \{D^* : D \in L\} \qquad (5.3)$$

$$\alpha^*(D^*) = -\alpha(D)^* \ . \qquad (5.4)$$

We must check that $(\mathcal{D}'^*, L^*, \alpha^*)$ defined by these formulas really is a Lax system

$$[D^*, \alpha^*(D^*)] = -[D^*, \alpha(D)^*]$$

$$= -[\alpha(D), D]^* \in L^* \quad . \tag{5.4}$$

<u>Definition</u>. The Lax system $(\mathcal{D}', L, \alpha)$ is said to be *self-adjoint* if it is equal to its own adjoint, i.e., if $\mathcal{D}'^* = \mathcal{D}$, $L^* = L$, and

$$\alpha^* = \alpha \tag{5.5}$$

<u>Theorem 5.1</u>. If each $D \in \mathcal{D}'$ and each $\ell \in L$ is self-adjoint, i.e., satisfies

$$D^* = D$$

$$\ell^* = D \quad ,$$

then "self-adjointness" for the Lax system means that $\alpha(D)$ is skew-adjoint for $D \in \mathcal{D}'$.

<u>Proof</u>. This follows from (5.4).

<u>Theorem 5.2</u>. If $t \to D(t)$ is a solution of the Lax equation

$$\frac{dD}{dt} = [D, \alpha(D)]$$

for the Lax system $(\mathcal{D}', L, \alpha)$, then

$$t \to D(t)^*$$

is a solution of the Lax equation for the adjoint system $(\mathcal{D}'^*, L^*, \alpha^*)$.

<u>Proof</u>.

$$\frac{dD(t)^*}{dt} = [D, \alpha(D)]^*$$

$$= [\alpha(D)^*, D^*]$$

$$= [-\alpha^*(D^*), D^*]$$

$$= [D^*, \alpha^*(D^*)] \quad \text{Q.E.D.}$$

The Muira transform now attempts to construct a self-adjoint Lax system from an arbitrary one. Start off with

Korteweg-De Vries

$$(\mathcal{D}', L, \alpha)$$

as a Lax system. Construct its adjoint

$$(\mathcal{D}'^*, L^*, \alpha^*)$$

Now, set

$$\mathcal{D}_M = \{DD^* : D \in \mathcal{D}'\} \qquad (5.6)$$

Suppose $t \to D(t)$ is a solution of the Lax equation for the Lax system $(\mathcal{D}', L, \alpha)$:

$$\frac{dD}{dt} = [D, \alpha(D)] \quad .$$

Set:

$$D_M(t) = D(t)D(t)^* \quad . \qquad (5.7)$$

Then,

$$\begin{aligned}
\frac{dD_M}{dt} &= \frac{dD}{dt} D^* + D \frac{dD^*}{dt} \\
&= [D, \alpha(D)]D^* + D[D, \alpha(D)]^* \\
&= [D, \alpha(D)]D^* + D[\alpha(D)^*, D^*] \\
&= [D, \alpha(D)]D^* + D[D^*, \alpha^*(D)] \qquad (5.8)
\end{aligned}$$

Let us try to find an

$$\alpha_M(D)$$

such that

$$\frac{dD_M}{dt} = [D_M, \alpha_M(D)] \qquad (5.9)$$

$$\alpha_M(D)^* = -\alpha_M(D) \quad . \qquad (5.10)$$

This requires that

$$\begin{aligned}
\frac{dD_M}{dt} &= [DD^*, \alpha_M(D)] \\
&= [D, \alpha_M(D)]D^* + D[D^*, \alpha_M(D)] \qquad (5.11)
\end{aligned}$$

Let us try

$$\alpha_M(D) = (\alpha(D) - \alpha(D)^*) \qquad (5.12)$$

Insert (5.12) into (5.11):

$$\frac{dD_M}{dt} = [D,(\alpha(D) - \alpha(D)^*]D^* + [D^*,(\alpha(D) - \alpha(D)^*]$$

$$= ([D,\alpha(D)]D^* - D[D^*,\alpha(D)^*]) + (-[D,\alpha(D)^*]D^* + D[D^*,\alpha(D)])$$

$$= \text{, using (5.8)}$$

$$\frac{dD_M}{dt} + D[D^*,\alpha(D)] + (D[D^*,\alpha(D)])^*$$

Hence, consistency requires that:

$$D[D^*,\alpha(D)] \text{ is skew-adjoint for } D \in \mathcal{D} \qquad (5.13)$$

Theorem 5.3. Suppose that condition (5.13) is satisfied. Then, for each solution $t \to D(t)$ of the Lax equation associated with the given Lax system $(\mathcal{D}', L, \alpha)$,

$$t \to D_M(t) = D(t)D(t)^*$$

is a solution of a Lax-type equation

$$\frac{dD_M}{dt} = [D_M, \alpha_M(D_M)] \qquad (5.14)$$

with

$$\alpha_M(D_M) = \alpha(D) - \alpha(D)^* \qquad (5.15)$$

6. THE MOYAL ALGEBRA

We will now introduce a more efficient way of carrying out these calculations, closely linked to the "symbolic" methods of 19th century invariant theory, on the one hand, and on the other to quantum mechanics.

Introduce two real variables p and q. Let $\mathcal{D}$ consist of the real valued C^∞ function

$$f(p,q)$$

of these variables, which is a polynomial in p.

Korteweg-De Vries

$$\partial_p f, \quad \partial_q f, \quad \text{or} \quad f_p, f_q$$

Set:

$$\Delta_n(f_1, f_2) = \sum_{j=0}^{\infty} (-1)^j \binom{n}{j} \partial_p^{n-j} \partial_q^j (f_1) \partial_q^{n-j} \partial_p^j (f_2) \tag{6.1}$$

This differential equation Δ_m is called the *m-th order transvection*. It is a differential operator which played a prominent role in 19th century invariant theory. Now, for $f_1, f_2 \in \mathcal{D}$, set:

$$f_1 \# f_2 = \sum_{m=0}^{\infty} \frac{1}{m!} \Delta_m(f_1, f_2) \tag{6.2}$$

Then,

$$[f_1, f_2] = f_1 \# f_2 - f_2 \# f_1$$

$$= \sum_{j=0}^{\infty} \frac{(-1)^j}{(2j+1)!} \Delta_{2j+1}(f_1, f_2) \tag{6.3}$$

Define an adjoint operation $*$ on $\mathcal{D}$ as follows:

$$f(p,q)^* = f(-p,q) \ . \tag{6.4}$$

The Moyal product was first introduced as the "missing link" between Poisson bracket and Weyl's method of quantization. Here, it has nothing (directly) to do with quantization, but is a very efficient algebraic tool to carry out the calculation required of the Lax theory.

Here is another algebraic form of the Moyal product, which was suggested to me by Nolan Wallach (private communication).

Let

$$\mathcal{D} \otimes \mathcal{D}$$

denote the tensor product of two copies $\mathcal{D}$ as real vector spaces. Let

$$m: \mathcal{D} \otimes \mathcal{D} \to \mathcal{D}$$

be the linear map defined as follows

$$m(f_1 \otimes f_2) = f_1 f_2 \ . \tag{6.5}$$

Let

$$\pi: \mathcal{D} \otimes \mathcal{D} \to \mathcal{D} \otimes \mathcal{D}$$

be the linear map defined as follows:

$$(f_1 \otimes f_2) = \partial_p(f_1) \otimes \partial_q(f_2) - \partial_q(f_1) \otimes \partial_p(f_1) \tag{6.6}$$

Then,

$$\Delta_n = m\pi^n . \tag{6.7}$$

(π^n is the n-th product of π, as a linear map.)

Using this form (which shows how the Moyal product is a generalization of Poisson bracket) it is quite easy to show that the Moyal product is associative.

7. THE FIRST ORDER LAX SYSTEM AND THE MOYAL ALGEBRA

Let $\mathscr{D}'$ be the functions of p and q of the following form:

$$f = p + a(q) \tag{7.1}$$

with $a(q)$ an arbitrary function of q. Let $\alpha(f)$ be a monic polynomial of degree n in p

$$\alpha(f) = p^n + \alpha_{n-1}(q)p^{n-1} + \cdots + \alpha_0(p)$$

$$\Delta_m(\alpha(f),f) = \partial_p^m(\alpha(f))\partial_q^m(a) - m\partial_p^{m-1}\partial_q(\alpha(f)) \tag{7.2}$$

Hence,

$$[\alpha(f),f] = \partial_p(\alpha(f))\partial_q(a) - \partial_q(\alpha(f)) + \partial_p^3(\alpha(f))\partial_q^3(a) \\ - 3\partial_p^2\partial_q(\alpha(f)) - \cdots \tag{7.3}$$

For example, let us work out the case $n = 3$ explicitly:

$$[\alpha(f),f] = (3p^2 + 2\alpha_2 p + \alpha_1)a_q - (\alpha_{2,q}p^2 - \alpha_{1,q}p - \alpha_{0,q}) \\ + 6a_{qqq} - 6\alpha_{2,q} \tag{7.4}$$

For the Lax condition, we want this to be of degree 0 in q. This requires that the terms in p^2 and p vanish, i.e., that the following conditions be satisfied:

$$3a_q - \alpha_{2,q} = 0 ,$$

or

$$3a - \alpha_2 = c_2 \;. \tag{7.5}$$

(c denotes constant.)

$$2\,_2a_q + \,_{1,q} = 0$$

or

$$2(3a - c_2)a_q + \alpha_{1,q} = 0$$

$$3a^2 - 2c_2 a + \alpha_1 = c_1 \;. \tag{7.6}$$

Chapter 9

THE GENERALIZED TODA LATTICES AS CAUCHY CHARACTERISTIC VECTOR FIELDS

1. INTRODUCTION

The Toda lattice appeared originally [1] as a mechanical system of particles on the line governed by a certain type of nearest neighbor interaction. With the work of Flaschka [2] a relation with Lie group theory, particularly earlier work by Arnold [3] on the rotating rigid body and its generalizations, came into the foreground; this relation has been extensively developed since, most notably in recent work by Kostant [4]. In previous work [5,6] I have developed certain relations between the Toda lattice and certain types of vector fields on Lie algebras. In this paper I will develop these relations in a more definitive form.

The basic idea of [5] was that the Toda lattice should most naturally be identified with the extremals of a left-invariant calculus of variations system on a Lie group [9]. Now, it is known since the work of Cartan [7] that such extremals can be described in an elegant differential-geometric way by what are called *Cauchy characteristic* curves of closed two-differential forms [8.19]. Thus, implicitly there is an identification of the Toda lattice equations with certain Cauchy characteristic equations. The aim of this paper is to bring this relation more into the foreground and to suggest certain generalizations of the Toda lattice which fit in very well with this identification. Notably, we extend the Toda lattice both in terms of Lie groups (unifying certain themes of [4] and [5]) and in terms of spaces with *absolute parallelism*. We also develop some underlying Lie algebra theory.

2. CAUCHY CHARACTERISTICS OF CLOSED TWO-DIFFERENTIAL FORMS

First, we shall review certain differential-geometric fundamentals [8]. Let M be a manifold and let ω be a closed two-differential form on M. If $v \in T(M)$ is a tangent vector to M at a point $p \in M$, the inner product or contraction of ω by v is denoted as $i(v)(\omega)$; it is a one-covector at p, i.e., an element of the dual space to the tangent space to M at p. Similarly, if V is a vector field on M, $i(V)(\omega)$ is defined as a one-differential form on M.

Definition. A tangent vector $v \in T(M)$ is said to be Cauchy characteristic for ω if $i(v)(\omega) = 0$. Similarly, a tangent vector field V is said to be Cauchy characteristic if $i(V)(\omega) = 0$.

The basic role that the Cauchy characteristics play in the calculus of variations and Hamilton-Jacobi theory was first pointed out by Cartan [7].

In this paper we shall work with a special choice of the manifold M and the closed two-differential form. Namely, suppose given the following data:

A vector space X and a manifold Y, such that $M = X \times Y$.

An absolute parallelism on Y defined by a basis θ^a, $1 \leq a, b \leq m$, of one-differential forms on Y.

A basis x_i, $1 \leq i, j \leq n$, of the linear functions on X.

Adopt the summation convention on the indices given above. In addition, suppose that $m > n$. Introduce the following additional indices and the summation convention on these indices:

$$n + 1 \leq u, v \leq m \ .$$

Let f^a_{bc} be the *structure functions* of the absolute parallelism, i.e., the functions on Y such that:

$$d\theta^a = f^a_{bc} \theta^b \wedge \theta^c \ .$$

Let G be the automorphism group of the absolute parallelism, i.e., the group of diffeomorphisms $g: Y \to Y$ such that:

$$g^*(\theta^a) = \theta^a \ .$$

It is known that G is a Lie group and that it acts simply on Y, i.e., the orbits of G can be identified with G itself. We shall suppose that the orbit space

$$Z = G \backslash Y$$

is a manifold and that the quotient map $Y \to Z$ is a submanifold map. The structure functions f^c_{ab} are constant on the orbits of G, hence are pullbacks under the quotient map of functions on Z. We shall make no notational distinction between these functions.

Now, set:

$$\omega = d(x_i \theta^i) \ .$$

Let us compute the Cauchy characteristic vectors of ω, using the relations given above.

Toda Lattices 127

$$\omega = dx_i \wedge \theta^i + x_i d\theta^i \ .$$

Our job is to put the differential form ω into its algebraic "canonical form". To do this, note that

$$q\omega = (dx_i - x_j f^j_{ai} \theta^a) \wedge \theta^i + x_i f^i_{uv} \theta^u \wedge \theta^v \ .$$

Set:

$$\alpha_i = dx_i - x_j f^j_{ai} \theta^a$$

$$\omega' = x_i f^i_{uv} \theta^u \wedge \theta^v \ .$$

Then, we have the definitive formula:

$$\omega = \alpha_i \wedge \theta^i + \omega' \ , \qquad (2.1)$$

where the differential forms on the right hand side of (2.1) are linearly independent. Thus, we have proved the following result:

<u>Theorem 2.1</u>. The Cauchy characteristic vector fields V of ω satisfy the following equation:

$$\begin{aligned} 0 &= i(V)(\alpha_i) \\ &= i(V)(\theta^i) \\ &= i(V)(\omega') \ . \end{aligned} \qquad (2.2)$$

<u>Corollary</u>. The dimension of the Cauchy characteristic tangent vectors to ω is equal to the dimension of the Cauchy characteristic tangent vectors to ω'.

We can now work out the equations for the Cauchy characteristic vector field V defined by relation (2.2) in more detail. First, let us work with the second relation on the right hand side of (2.2). Suppose that we impose the following relations:

$$\theta^u(V) = h^u \ , \qquad (2.3)$$

where the h^u are functions of the x's and the f's. They must then satisfy the following condition:

$$x_i f^i_{uv} h^u = 0 \ . \qquad (2.4)$$

With this choice for these functions, we see that V is completely determined by the first part of relations (2.2):

$$\theta^i(V) = 0,$$

$$V(x_i) = x_j f^j_{ui} \theta^u(V)$$

$$= x_j f^j_{ui} h^u \qquad (2.5)$$

Here is an important geometric property of the Cauchy characteristic vector fields of this form that is proved by the relations described above:

<u>Theorem 2.2</u>. Let V be a Cauchy characteristic vector field of ω given by relations (2.3)-(2.5). Let $\phi: M \to X \times Z$ be the map which sends the point $(x,y) \in X \times Y = M$ into (x,z), where z is the orbit of G acting on Y, which contains the point y. Then, the vector field V projects into $X \times Z$, i.e., there is a vector field V' on $X \times Z$ such that ϕ sends each orbit curve of V into an orbit curve of V'.

In practice, we often start off with V' and construct V. Notice that it is essentially this construction that defines the "symplectic" structure for the orbit curves of V'; the situation is simplest in case Z reduces to a point, i.e., in case the f's are constants. This means that Y is equal to the Lie group G itself, with G acting by left translation. In this case, we shall see that the equations for the orbit curves of V' are the differential equations for the Today lattices and their generalizations.

3. SPECIALIZATION TO THE ABSOLUTE PARALLELISM DEFINED BY THE LEFT INVARIANT DIFFERENTIAL FORMS ON A LIE GROUP

Let us now apply Theorem 2.1 to the case that the θ^a are the left-invariant Cartan-Maurer forms for a Lie group G and the x_a are the dual linear coordinates for the dual vector space $\underline{G}^d$ of the Lie algebra of G. Consider X as $\underline{J} \times G$, where $\underline{J}$ is the linear subspace of $\underline{G}^d$ defined by the relations $x_u = 0$. Let $\underline{J}'$ be the orthogonal complement of $\underline{J}$ in $\underline{G}$.

<u>Theorem 3.1</u>. At a point x of $\underline{J}$, the dimension of the Cauchy characteristic vectors of the two-differential form ω' is equal to the dimension of the characteristic subspace of the skew-symmetric bilinear form

$$(J_1, J_2) \to x([J_1, J_2])$$

on $\underline{J}'$. An element $J_1 \in \underline{J}'$ is Cauchy characteristic for the form ω' at x if and only if the following condition is satisfied:

$$\text{co Ad } J_1(x) \in \underline{J}. \qquad (3.1)$$

Toda Lattices 129

Remark. "coAd" means the dual of the adjoint representation of the Lie algebra $\underline{G}$, i.e.,

$$\text{coAd}(A)(x)(B) = -x([A,B])$$

$$\text{for } x \in \underline{G}^d; \quad A,B \in \underline{G}.$$

4. FLASCHKA VECTOR FIELDS ON VECTOR SPACES

The conditions found for Cauchy characteristic vector fields in previous sections are sufficiently interesting and important that it is worth our while to pause and make some general definitions. Let $\underline{G}$ be a real Lie algebra. Let $\underline{G}^d$ be its dual space. Let X be a linear subspace of $\underline{G}^d$. Let X' be the orthogonal complement of X in $\underline{G}$, i.e., the set of elements $A \in \underline{G}$ such that $X(A) = 0$.

Definition. A *Flaschka map* for the vector space X is a linear map $F: X \to X'$ such that the following condition is satisfied:

$$\text{coAd}(F(x))(x) \in X \qquad (4.1)$$

$$\text{for all } x \in X .$$

With condition (4.1) satisfied, we can define a vector field V' on X considered as a manifold. Since X is a vector space, a "vector field" is just a map $X \to X$. Let us then set:

$$V'(x) = \text{coAd}(F(x))(x) \qquad (4.2)$$

$$\text{for all } x \in X .$$

Condition (4.1), of course, is precisely that which guarantees that V' defined by formula (4.2) is indeed a well-defined vector field on X. We shall call V' a *Flaschka vector field*, since we shall see that Flaschka's work on the Toda lattice fits into this framework very naturally.

The orbit curves of the vector fields V' are then the solutions $t \to x(t)$ of the following differential equations:

$$\frac{dx}{dt} = \text{coAd}(F(x(t))(x(t)) . \qquad (4.3)$$

Conversely, if we start off with the nonlinear differential equations (4.3) (which would be the normal thing to do), we would have the following properties:

Theorem 4.1. Consider the system of ordinary differential equations defined by relations (4.3), where F is a Flaschka map, as defined above. Construct the manifold M as $X \times \underline{G}$, and construct the two-differential form ω on M

as in the previous section. Then, the solution curves of Equation (4.3) are the projection in X of Cauchy characteristic curves of ω. In particular, this imposes in a natural way a symplectic structure on the space of solutions of (4.3).

So far, we have been working with an arbitrary Lie algebra $\underline{G}$. For a reductive Lie algebra, i.e., one for which $\underline{G}$ and $\underline{G}^d$ are naturally isomorphic, the formulas can be readily recast so as to be closer to those in the applied mathematics literature.

5. THE FLASCHKA MAPS AND VECTOR FIELDS FOR REDUCTIVE LIE ALGEBRAS

Let us now make the assumption that there is a non-degenerate symmetric bilinear form $\underline{B}$ on the (finite dimensional) Lie algebra $\underline{G}$, which is invariant under the adjoint representation of $\underline{G}$ on itself. $\underline{B}$ sets up an isomorphism between $\underline{G}$ and its dual space $\underline{G}^d$. Let $\underline{J}$ be a linear subspace of $\underline{G}^d$; under this identification of $\underline{G}^d$ with $\underline{G}$, $\underline{J}$ is identified with a linear subspace of $\underline{G}$. The orthogonal subspace that we denoted as $\underline{J}'$ is then identified with the orthogonal complement of $\underline{J}$ with respect to the form $\underline{B}$, i.e.,

$$\underline{J}' = \{A \in \underline{J}: \underline{B}(A,\underline{J}) = 0\} \quad .$$

A "Flaschka·map" is then a linear map

$$B: \underline{J} \to \underline{J}'$$

such that the vector field

$$V(A) = [B(A),A]$$

is tangent to $\underline{J}$.

Ways of constructing such maps (abstracted from the original Toda lattice work) were presented in [5]. Now we shall present this work from a more unified algebraic point of view.

6. EULER-ARNOLD VECTOR FIELDS ON LIE ALGEBRA

In this section we shall review certain methods by means of which differential equations may be defined that have some of the properties suggested by the classical rotating rigid body and the recent work on Toda lattices. These differential equations are essentially defined by means of certain types of vector fields on Lie algebras.

Toda Lattices   131

Let $\underline{G}$ be a Lie algebra. Since $\underline{G}$ is itself a vector space, a vector field (in the sense of manifold theory) is a map $V: \underline{G} \to \underline{G}$. The orbits or integral curves of such a vector field are the curves $t \to A(t)$ in $\underline{G}$ such that

$$\frac{dA}{dt} = V(A(t)) \ . \tag{6.1}$$

Such a vector field will be said to be of Euler-Arnold type if it is of the following form

$$V(A) = [B(A), A] \ , \tag{6.2}$$

where B is some map from $\underline{G}$ to $\underline{G}$.

In [5,6] I have shown how the construction of such a B and V given in the original Flaschka work on the Toda lattice may be described in terms of certain "graded" algebra structures on Lie algebras. Here is a more systematic study of this material.

7. JACOBI TRIPLES OF LIE ALGEBRAS

Let $\underline{G}$ be a real Lie algebra. A triple $(\underline{G}^+, \underline{G}^0, \underline{G}^-)$ of linear subspaces of $\underline{G}$ is said to define a *Jacobi triple* if the following conditions are satisfied:

$$\begin{aligned} [\underline{G}^0, \underline{G}^0] &\subset \underline{G}^0 \\ [\underline{G}^0, \underline{G}^{+,-}] &\subset \underline{G}^{+,-} \\ [\underline{G}^+, \underline{G}^-] &\subset \underline{G}^0 \ . \end{aligned} \tag{7.1}$$

The linear subspace

$$\underline{J} = \underline{G}^+ + \underline{G}^- + \underline{G}^0$$

spanned by these three subspaces is called the *Jacobi subspace* of $\underline{G}$ associated with the Jacobi triple.

Example. JACOBI MATRICES

Let V be a vector space and let $V_1, \ldots, V_n$ be linear subspaces of V. Let $\underline{G}$ be the Lie algebra (under commutator) of linear maps $V \to V$. Set:

$$\underline{G}^0 = \{A \underline{G}: A(B_i) \subset V_i, \ i = 1, \ldots, n\}$$

$$\underline{G}^{+,-} = \{A \underline{G}: A(V_i) \subset V_{i+,-1}, \ i = 1, \ldots, n\} \ .$$

(Thus, $\underline{G}^{+,-}$ are the "shift up" and "shift down" operators.) It is obvious that the commutation relations (7.1) are satisfied. If the $V_1,\ldots,V_n$ are one-dimensional linearly independent subspaces which span $\bar{V}$, and if a basis is chosen for V consisting of vectors from these subspaces, it is clear that the operators in $\underline{J}$ are represented by classical Jacobi matrices, i.e., $n \times n$ matrices with nonzero entries only on the diagonal, sub, and super diagonal lines. It is the point of view developed in [5] that the corresponding gradation of $\underline{G}$ gives rise to the Toda lattice phenomenon.

8. SIMPLE ROOT SYSTEMS FOR SIMPLE LIE ALGEBRAS AND JACOBI TRIPLES

It is known that the "Toda lattice" models are closely related to the algebraic properties of Lie algebras, particularly the properties of the simple root systems of semisimple Lie algebras. In this section I will show how the simple root systems generate Jacobi triples.

Let $\underline{G}$ be a finite dimensional simple Lie algebra. For the moment, assume that the field of scalars is the complex numbers. We shall return to the case of the real numbers later on. It will, of course, be assumed that the reader knows basic semisimple Lie algebra theory.

Let $\underline{C}$ be a Cartan subalgebra of $\underline{G}$. Ad $\underline{C}$, acting in $\underline{C}$, is then completely reducible. The nonzero eigenvalues of Ad $\underline{C}$, considered as linear forms on $\underline{C}$, are called the roots of the Lie algebra. Let $r = \dim \underline{C}$. (r is the rank of the Lie algebra $\underline{G}$.) A set $\lambda_1,\ldots,\lambda_r$ of roots is said to define a simple root system if the following conditions are satisfied:

> Each root $q1$ can be written as a linear combination of the $\lambda_1,\ldots,\lambda_r$ with coefficients that are integers, and that are simultaneously all non-negative or non-positive.
>
> $-\lambda_1,\ldots,-\lambda_r$ are roots.

That such simple root systems exist and serve to determine the isomorphism class of the Lie algebra is a well-known fact of Lie algebra theory.

Fix such a simple root system. Let (A_i), $i = 1,\ldots,r$, be a collection of root elements, i.e., vectors of $\underline{G}$ such that

$$[A,A_i] = q1_i(A)A_i \quad \text{(no summation)}$$

for all $A \in \underline{G}$.

(It is known from Lie algebra theory that there is, up to a constant multiple, just one such root vector.)

Since $\lambda_i - \lambda_j$ is not a root, for $i \neq j$ we have:

$$[A_i, A_j] = 0, \quad \text{for } i \neq j.$$

Toda Lattices 133

Similarly, let A_{-i} be a root vector for the root $-\lambda_i$. We then have

$$[A_{-i}, A_{-j}] = 0, \quad \text{for } i \neq j.$$

Also,

$$[A_i, A_{-i}] \in \underline{C}.$$

Thus, if we let $\underline{G}^+$ ($\underline{G}^-$) be the linear subspace spanned by the A_i (A_{-i}), and let $\underline{G}^0$ be $\underline{C}$, we see that $(\underline{G}^0, \underline{G}^+, \underline{G}^-)$ forms a Jacobi triple.

So far, $\underline{G}$ has been a complex Lie algebra. We can form a real subalgebra $\underline{G}'$ as the Lie subalgebra generated by the A_i, A_{-i}. $(\underline{G}^0, \underline{G}^+, \underline{G}^-)$ forms a Jacobi triple in $\underline{G}'$.

9. EULER-ARNOLD VECTOR FIELDS THAT ARE TANGENT TO THE JACOBI SUBSPACES

Let $\underline{G}$ be a real Lie algebra and let $(\underline{G}^+, \underline{G}^-, \underline{G}^0)$ be a Jacobi triple of linear subspaces. Let $\underline{J}$ be the associated Jacobi subspace of $\underline{G}$. Let $B: \underline{G} \to \underline{G}$ be a map, and set:

$$V(A) = [B(A), A]$$

for $A \in \underline{G}$.

Consider V as a vector field on $\underline{G}$. We ask:

When is V tangent to the linear subspace $\underline{J}$?

To answer this question, suppose that $B = B^+ + B^0 + B^-$, where

$$B^{+,-,0}(\underline{J}) \subset \underline{G}^{+,-,0}.$$

If $A = A^+ + A^- + A^0 \in \underline{J}$, then

$$V(A) = [B^+(A) + B^0(A) + B^-(A), A^+ + A^0 + A^0].$$

Let us now suppose that the following conditions are satisfied:

$$\begin{aligned}[B^+(A), A^+] &= 0 \\ [B^-(A), A^-] &= 0.\end{aligned} \quad (9.1)$$

<u>Theorem 9.1</u>. If conditions (9.1) are satisfied, then the Euler-Arnold vector field $V(A) = [B(A), A]$ is tangent to the Jacobi subspace $\underline{J}$.

<u>Proof</u>. Condition (9.1) implies that $V(\underline{J}) \subset \underline{J}$, which is the condition that $\underline{J}$ be tangent to $\underline{J}$.

10. JACOBI TRIPLES DEFINED BY AUTOMORPHISMS OF LIE ALGEBRAS

Following a suggestion by Victor Kac (private communication), we shall now show how Jacobi triples may be defined by automorphisms of Lie algebras. Let $\underline{G}$ be a finite dimensional Lie algebra with the complex numbers as field of scalars. Let $\sigma: \underline{G} \to \underline{G}$ be an automorphism of this Lie algebra. Let $\lambda \in C$ be an eigenvalue of . Set:

$$\underline{G}^0 = \{A \in \underline{G}: \sigma(A) = A\}$$

$$\underline{G}^+ = \{A \in \underline{G}: \sigma(A) = \lambda A\}$$

$$\underline{G}^- = \{A \in \underline{G}: \sigma(A) = \lambda^{-1} A\} \quad .$$

These three subspaces $(\underline{G}^0, \underline{G}^+, \underline{G}^-)$ then define a Jacobi triple. This way of defining Jacobi triples is especially natural because automorphisms are classifiable if $\underline{G}$ is a semisimple Lie algebra. The case where the automorphism is of finite order would be or particular interest, since Kac has classified them. (See Section 5, Chapter X of [22]. It is also shown in this reference how such automorphisms are related to graded structures on Lie algebras.)

As an illustration, let us construct the automorphism which gives rise to the classical Jacobi matrices. Let V be a finite dimensional complex vector space. Let

$$V = V_1 + \cdots + V_n$$

be a direct sum decomposition of V as a direct sum of linear subspaces. Let   be a primitive n-th root of unity, i.e.,

$$\lambda^n = 1 \quad ,$$

but $\lambda^j \neq 1$ for $j = 2, \ldots, n-1$.

Set

$$\sigma(v) = \lambda^j v$$

$$\text{for } v \in V_j, \quad j = 1, \ldots, n \quad .$$

Let $\underline{G}$ be $L(V,V)$, the Lie algebra of all linear maps: $V \to V$. qs defines a linear map (also denoted as σ) of $\underline{G} \to \underline{G}$:

$$\sigma(A) = \sigma A \sigma^{-1}$$

for $A \in \underline{G}$.

Set

Toda Lattices

$$\underline{G}^0 = \{A \in \underline{G}: \sigma(A) = A\}$$

$$\underline{G}^+ = \{A \in \underline{G}: \sigma(A) = \lambda A\}$$

$$\underline{G}^- = \{A \in \underline{G}: \sigma(A) = \lambda^{-1} A\}$$

We see that

$$\underline{G}^0(V_j) \subset V_j, \quad \text{for } j = 1,\ldots,n.$$

$$\underline{G}^+(V_j) \subset V_{j+1}$$

$$\underline{G}^-(V_j) \subset V_{j-1}.$$

These relations imply that $\underline{J} = \underline{G}^0 + \underline{G}^+ + \underline{G}^-$ is a linear subspace of $\underline{G}$, which is represented by Jacobi matrices in the classical sense, when a basis is chosen for V consisting of elements in the subspaces V_j.

11. FLASCHKA MAPS CONSTRUCTED FROM JACOBI TRIPLES

Suppose now that $\underline{G}$ is a reductive Lie algebra. Notice that the "Euler-Arnold" objects differ from the "Flaschka" objects only in that the values of the former type of map do not necessarily lie in the appropriate orthogonal complement. In this section we shall see what sort of compatibility condition between the graded structures and the Ad G-invariant symmetric bilinear form $\underline{B}: \underline{G} \times \underline{G} \to R$ must be imposed in order to assure that the "Flaschka" conditions are to be satisfied.

Suppose that $(\underline{J}^-, \underline{J}^+, \underline{J}^0)$ is a Jacobi triple of linear subspaces of the Lie algebra $\underline{G}$. In addition, let us suppose that $T: \underline{G} \to \underline{G}$ is a linear map such that the following condition is satisfied:

$$T([A_1, A_2]) = [T(A_2), T(A_1)] \tag{11.1}$$

$$\underline{B}(T(A_1), T(A_2)) = \underline{B}(A_1, A_2) \tag{11.2}$$

for $A_1, A_2 \in \underline{G}$.

$$T(T(A)) = A$$

for $A \in \underline{G}$ \hfill (11.3)

<u>Remark</u>. Notice that, if $\underline{G}$ is a Lie algebra of matrices, then $T(A)$ = transpose of A as a matrix will have these properties.

Let us now suppose that the Jacobi triple, the T-operation and the form $\underline{B}$, satisfy the following conditions:

$$T(\underline{J}^{\pm}) = \underline{J}^{\mp} \tag{11.4}$$

$$T(\underline{J}^{0}) = \underline{J}^{0} \tag{11.5}$$

$$\underline{B}(\underline{J}^{\pm}, \underline{J}^{\pm}) = 0 \tag{11.6}$$

$$\underline{B}(\underline{J}^{\pm}, \underline{J}^{0}) = 0 \;. \tag{11.7}$$

Let us now define the linear subspace $\underline{J}$ of $\underline{G}$ as the set of all elements of $\underline{G}$ of the following form:

$$J = J^{-} + J^{0} + T(J^{-}) \;, \tag{11.8}$$

where J^{-}, J^{0} are arbitrary elements of $\underline{J}^{-}$, $\underline{J}^{0}$. Define the linear map $B: \underline{J} \to \underline{G}$ by the following formula:

$$B(J) = J^{-} - T(J^{-}) \;. \tag{11.9}$$

<u>Theorem 11.1</u>. The map defined by formula (11.9) has the Flaschka property, i.e., the following conditions are satisfied:

$$[B(J), J] \in \underline{J} \;, \quad \text{and}$$

$$B(J) \in \underline{J}' \;,$$

for $J \in \underline{J}$, and with $\underline{J}'$ the orthogonal complement of $\underline{J}$ with respect to the form $\underline{B}$.

The <u>proof</u> of these statements is now an easy consequence of our assumptions.

12. THE "INTEGRABILITY" PROPERTIES OF FLASCHKA VECTOR FIELDS

Let us now return to the main theme of this paper, the study of Cauchy characteristic vector fields. Let $\underline{G}$ be a Lie algebra, and let X be a linear subspace of the dual space $\underline{G}^{\overline{d}}$. Let $B: X \to \underline{G}$ be a Flaschka map, and let

$$V'(x) = \text{coAd}\;(B(x))(x)$$

be the corresponding Flaschka vector field. Set:

$$M = X \times G \;.$$

Toda Lattices

We have constructed a two-differential form ω on M and a vector field V on M with the following properties:

$$d\omega = 0$$

$$i(V)(\omega) = 0$$

$$V \text{ projects into } V'.$$

Let $S(\underline{G})$ be the symmetric tensor algebra of $\underline{G}$. It has a Lie algebra structure. ($S(\underline{G})$ is the graded vector space associated with the natural filtration on the universal enveloping algebra $U(\underline{G})$ of the Lie algebra $\underline{G}$. The natural Lie algebra structure on $U(\underline{G})$ defines the Lie algebra structure on $S(\underline{G})$.) Each element of $\underline{G}$ defines a linear function on X; the polynomials in these linear functions are isomorphic to $S(\underline{G})$. Thus, there is defined a Lie algebra structure on the space of polynomial functions on $\underline{G}^d$. Let us suppose that V', as a vector field on X, is the restriction of a vector field V'' on $\underline{G}$ and that there is a polynomial function h on X such that:

$$V''(f) = \{h, f\}, \qquad (12.1)$$

for all $f \in S(\underline{G})$, where $\{\,,\,\}$ denotes the Lie algebra structure on the polynomial functions on X.

Let $C(\underline{G})$ be the center of the Lie algebra $S(\underline{G})$. With these assumptions, we see that $C(\underline{G})$ will define a vector space of functions on X with the following property:

$$\text{Each } f \in C(\underline{G}) \text{ is a constant of motion for the orbit curves of } V'. \qquad (12.2)$$

These constants of motion are, in a sense, given to us "free"; we are interested in seeing how they can be used to linearize the differential equations for the orbit curves of V'. To do this, let us pull everything back to M.

On $M = X \times G$, let, as before, x_i be a basis for linear functions on X and let θ^a be a basis for the left-invariant differential forms on G. Set:

$$\omega = d(x_i \theta^i).$$

Also, suppose that V is a vector field such that:

$$i(V)(\omega) = 0$$

and

$$V(x_i) = \text{functions of the } x\text{'s alone.}$$

Suppose also that $f_1,\ldots,f_r$ are elements of $C(\underline{G})$, considered as functions of the x's. For each choice of $\underline{c} = (c_1,\ldots,c_r)$ R, let:

$$M(\underline{c}) = \{p \in M: f_1(p) = c_1;\ldots;f_r(p) = c_r\} \quad .$$

Then, restrict ω to the submanifold $M(\underline{c})$. Suppose that the rank of ω, so restricted, is equal to s. Then, putting this restricted form into its "Darboux" canonical form will provide 2s independent functions $z_1,\ldots,z_{2s}$ and functions $g_1,\ldots,g_r$ such that the following conditions are satisfied:

$$\omega = dz_1 \wedge dz_2 + \cdots + dz_{2s-1} \wedge dz_{2s} + dg_1 \wedge df_1 + \cdots \quad .$$

Now, since the f's are constants of motion for the vector field V, the z's have the property that V applied to them are again functions of the z's. This, together with the condition that V be Cauchy characteristic for ω, implies that the following conditions are satisfied:

$$0 = d(V(g_1)) = \cdots + d(V(g_r)) \quad .$$

Thus, we see that, in the language of classical Hamilton-Jacobi theory, the g's will be the "angle" variables corresponding to the "action" variables f. In favorable cases, these functions, when projected down to X, will suffice for the "integration" of the vector field V'.

References

1. R. Hermann, Some differential geometric aspects of the Lagrange variational problem, *Illinois J. Math.* 6, 634-673 (1962).

2. R. Hermann, E. Cartan's geometric theory of partial differential equations, *Advances in Math.* 1, 265-317 (1965).

3. V. Arnold, Sur la geometrie differentielle des groupes de Lie de dimension infinie et ses applications a l'hydrodynamique des fluids parfaits, *Ann. Inst. Fourier Grenoble* 16:1, 319-361 (1966).

4. B. Kostant, The solution to a generalized Toda lattice and representation theory, MIT Preprint, 1979.

5. R. Hermann, *Toda Lattices, Cosymplectic Manifolds, Bäcklund Transformations and Kinks, Part A*, (Interdisciplinary Mathematics, Vol. 15), Math Sci Press, Brookline, MA 1977.

6. E. Cartan, *Lecons sur les Invariants Integraux*, Herman, Paris, 1922.

7. R. Hermann, *Differential Geometry and the Calculus of Variations*, Academic Press, New York, 1969. Second Edition, Math Sci Press, 1977.

8. R. Hermann, Left invariant geodesics and classical mechanics on manifolds, *J. Math. Phys.* 13, 460 (1972).

9. R. Hermann, Spectrum-generating algebras in classical mechanics, I and II, *J. Math. Phys.* 13, 833, 878 (1972).

10. R. Hermann, *Geometry, Physics and Systems*, Marcel Dekker, New York, 1973.

11. R. Hermann, Geodesics of singular Riemannian metrics, *Bull. Am. Math. Soc.* 79, 780-782 (1973).

12. O.I. Bogoyavlensky, On perturbations of the periodic Toda lattice, *Comm. Math. Phys.* 51, 201-209 (1976).

13. J. Moser, Finitely many mass points on the line under the influence of an exponential potential-an integrable system, *Battelle Rencontres*, 1974.

14. M.A. Olshanetzki and A.M. Perelmonov, Completely integrable Hamiltonian systems connected with semi-simple Lie algebras, *Inv. Math.* 37, 93-108 (1976).

15. M.A. Olshanetzki and A.M. Perelomov, Explicit solutions of the generalized Toda models

16. P. van Meerbecke, The spectrum of the Jacobi matrices, *Inv. Math.*

17. D. Kazhdan, B. Kostant, and S. Sternberg, Hamiltonian group actions and dynamical systems of Calogero type, *Comm. Pure and Appl. Math.* 31, 481-508 (1978).

18. M. Toda, Studies of a nonlinear lattice, *Phys. Lett.* 8, 1-125 (1975).

19. H. Flashka, On the Toda lattice, I, *Phys. Rev.* B9, 1924-1925 (1974); II, *Prog. Theo. Phys.* 51, 703-716 (1974).

20. S. Helgason, *Differential Geometry, Lie Groups and Symmetric Spaces*, Academic Press, New York, 1979.

21. V. Arnold, *Mathematical Methods of Classical Mechanics*, MIR, Moscow, 1975 (Springe- Graduate Texts in Math. No. 60, Springer-Verlag, New York).

22. R. Hermann, *Lie Algebras and Quantum Mechanics*, W.A. Benjamin, New York, 1970.

PART III

MECHANICS AND CONTROL

Chapter 10

TOWARD THE GEOMETRIC UNIFICATION OF OPTIMAL CONTROL THEORY, THE CALCULUS OF VARIATIONS, AND ANALYTICAL MECHANICS

1. INTRODUCTION

Although there are close relations between control theory and analytical mechanics as mathematical disciplines, they have developed in virtually complete isolation from each other. The resemblance between them is especially close when they are studied with the tools bequeathed to us by Elie Cartan--manifold and fiber bundle theory, differential forms and exterior differential systems. My purpose in this paper is to begin such a unified study.

There is already much in my own work about this interrelation; I hope to present it in a more unified and systematic way here, and to indicate certain new and interesting lines of research.

Of course, to a historian of science there would be nothing surprising about a close link between optimal control theory and mechanics--both grew out of the part of classical mathematics called "The Calculus of Variations". There is also much of potential interest to the current science and technology in understanding more clearly the relations between these subjects. After all, many problems encountered in today's world involve intermingled aspects of "mechanics" and "control".

In this paper I will use a very general version of the classical calculus of variations based on the use of differential forms. This will cover what are classically called "simple" and "multiple" integral calculus of variations problems, and much besides.

2. NOTATION AND SOME BASIC CONCEPTS OF DIFFERENTIAL GEOMETRY

One of the main themes of modern differential geometry is the translation of geometry into algebra, in much the same spirit as the Cartesian "analytical geometry" translates Euclidean geometry into polynomial algebra. Of course the context is much wider, and one must understand algebra in a broader sense, including Lie and Grassmann algebra theory and the quasi-algebrazation of the theory of differential equations developed by Elie Cartan and now called *the theory of exterior differential systems*.

Let us first recall some basic algebraic concepts. Let K be a field. (The real or complex numbers would suffice for our purposes.) An algebra $\mathcal{A}$ is defined as a set $\mathcal{A}$, with elements denoted as $A, A_1, A_2, \ldots$, together with a K-vector space structure on $\mathcal{A}$, i.e., an addition operation

$$A_1, A_2 \to A_1 + A_2 ,$$

and a K-bilinear map

$$(A_1, A_2) \to A_1 A_2 ,$$

called *multiplication*.

A K-vector space $\mathcal{A}$ has a Z^+-*graded structure* if it is the direct sum

$$A = \mathcal{A}^0 \oplus \mathcal{A}^1 \oplus \cdots$$

of linear subspaces $\mathcal{A}^0, \mathcal{A}^1, \ldots$ labelled by the non-negative integer-.

Definition. A *Grassmann algebra* (over the scalar field K) is defined as a K-vector space $\mathcal{A}$, together with a Z^+-graded structure and an algebraic structure on $\mathcal{A}$ satisfying the following conditions:

$$\mathcal{A}^m \mathcal{A}^n \subset \mathcal{A}^{m+n} \tag{2.1}$$

for integers $m, n \geq 0$.

$$A_1(A_2 A_3) = (A_1 A_2) A_3 \tag{2.2}$$

for $A_1, A_2, A_3 \in \mathcal{A}$,

i.e., the algebraic structure is associative.

$$A_1 A_2 = (-1)^{mn} A_2 A_1 \tag{2.3}$$

for $A_1 \in \mathcal{A}^m$; $A_2 \in \mathcal{A}^n$.

The Grassmann algebras we consider in this paper have an identity element, which we denote as "1", and which lie in $\mathcal{A}^0$. Thus, the field K is embedded as a subspace of $\mathcal{A}^0$.

Notice that $\mathcal{A}^0$ is a subalgebra of $\mathcal{A}$, and that:

$$\mathcal{A}^0 \mathcal{A}^n = \mathcal{A}^n \mathcal{A}^0$$

$$= \mathcal{A}^n .$$

Thus, each graded subspace $\mathcal{A}^n$ is a module over the algebra $\mathcal{A}^0$.

Definition. A *differential Grassmann algebra* is defined as a Grassmann algebra $\{\mathcal{A}, \mathcal{A}^n, n \geq 0\}$, together with a K-linear map

$$d: \mathcal{A} \to \mathcal{A}$$

which satisfies the following conditions:

$$d\mathcal{A}^n \subset \mathcal{A}^{n+1}, \tag{2.4}$$

$$n = 0, 1, \ldots$$

$$d^2 = 0. \tag{2.5}$$

$$d(A_1 A_2) = dA_1 A_2 + (-1)^m A_1 dA_2 \tag{2.6}$$

for $A_1 \in \mathcal{A}^m$, $A_2 \in \mathcal{A}^n$.

Let us suppose given one or more such differential Grassmann algebras, and define several auxilliary algebraic concepts. A *derivation* (geometrically a *vector field*) is a K-linear map

$$V: \mathcal{A} \to \mathcal{A}$$

such that:

$$V(\mathcal{A}^m) \subset \mathcal{A}^m \qquad (2.7)$$

for $m \geq 0$.

$$V(A_1 A_2) = V(A_1)A_2 + A_1 V(A_2) \qquad (2.8)$$

for $A_1, A_2 \in \mathcal{A}$.

The derivations form a Lie algebra under commutation:

$$(V_1, V_2) \to [V_1, V_2] \equiv V_1 V_2 - V_2 V_k \qquad (2.9)$$

A *contraction* is a K-linear mapping

$$W: \mathcal{A} \to \mathcal{A}$$

such that:

$$W(A^m) \subset A^{m-1} \qquad (2.10)$$

$$W(A_1 A_2) = W(A_1)A_2 + (-1)^m A_1 W(A_2) \qquad (2.11)$$

for $A_1 \in \mathcal{A}^m$, $A_2 \in \mathcal{A}$.

A homomorphism between two differential Grassmann algebras $(\mathcal{A}, d)$ and $(\mathcal{A}', d')$ is a K-linear mapping

$$\phi: \mathcal{A} \to \mathcal{A}'$$

such that

$$\phi(A^m) \subset A^m$$

for $m \geq 0$.

$$\phi d = d' \phi .$$

Thus, we see that the differential Grassmann algebras form a convenient "category" of algebraic objects. One can assign to each finite dimensional differentiable manifold X a differential Grassmann algebra $\mathcal{D}(X)$, the *differential forms*, such that $\mathcal{D}^0(X)$ is the algebra (under point-wise multiplication) of infinitely differentiable real-valued functions on X.

Geometric Unification

$\mathcal{D}(X)$ is generated as a differential Grassmann algebra by $\mathcal{D}^0(X)$. The precise definition and basic properties of "differentiable manifolds" and the detailed properties of the differential forms are covered in the treatises on differential geometry written since 1960; it must be assumed that the reader is somewhat familiar with this background material. The elements of $\mathcal{D}(X)$ are denoted by Greek letters, e.g., $\omega, \theta, \eta, \ldots,$. The algebra structure in $\mathcal{D}(X)$ is called *exterior multiplication*, and usually is denoted by a special symbol. For typographical convenience, we shall denote this product by juxtaposition. (Of course, the reader must keep in mind, when doing computations, that the product is not commutative.) The differential operation d is called *exterior derivative*.

Thus, we assign to each manifold X a set of differential forms $\{\omega\}$, satisfying algebraic and differential rules. To each differential map $\phi: X \to Y$ between manifolds there is a "pull-back" Grassmann algebra homomorphism

$$\phi^*: \mathcal{D}(Y) \to \mathcal{D}(X) \quad .$$

Each vector field V (defined as a derivation of $\mathcal{D}(X)$) is the "infinitesimal generator" of a "one parameter local group of diffeomorphisms of X", with the Lie algebra structure related algebraically to the group structure (i.e., composition) of the group of diffeomorphisms. We denote the set of vector fields as $\mathcal{V}(X)$. A special feature is that to each $V \in \mathcal{V}(X)$, one can assign a contraction operation

$$i(V): \mathcal{D}(X) \to \mathcal{D}(X) \quad ,$$

such that:

$$i(V_1)i(V_2) = -i(V_2)i(V_1) \qquad (2.12)$$

$$\text{for } V_1, V_2 \in \mathcal{V}(X) \quad .$$

This enables each $\omega \in D^m(X)$ to be identified with a map

$$(V_1, \ldots, V_m) \to i(V_m) \cdots i(V_1)(\ )$$

if

$$\mathcal{V}(X) \times \cdots \times \mathcal{V}(X) \to \mathcal{D}^0(X) \quad ,$$

which is $D^0(X)$-multilinear and skew-symmetric. This enables ω to be identified with the geometric objects which are called "skew-symmetric, covariant tensor fields" in classical tensor analysis. The following identity relates contraction and exterior derivative, and plays the basic role in the calculus of variations:

$$V(\omega) = d(i(V)(\omega)) + i(V)(d(\omega)) \qquad (2.13)$$

for $V \in \mathscr{V}(X)$ and $\omega \in \mathscr{D}(X)$.

Let us now abstract from E. Cartan's theory of exterior differential systems certain algebraic notions. Let $\mathscr{A}$ be a differential Grassmann algebra. A *Grassmann ideal* $\mathscr{I}$ is a linear subspace of $\mathscr{A}$, which is an ideal in the algebraic sense, i.e.,

$$A \mathscr{I} \subset I$$
$$\mathscr{I} A \subset I \qquad (2.14)$$

for all $A \in \mathscr{A}$.

<u>Definition</u>. An *exterior differential system* $\mathscr{E}$ is a Grassmann ideal which also satisfies:

$$d\mathscr{E} \subset \mathscr{E}.$$

<u>Remark</u>. Alternately, this is called a *differential ideal*.

Giving an exterior differential system $\mathscr{E}$ as a subset of the differential Grassmann algebra $\mathscr{D}(X)$ associated with a manifold X defines a family of submanifolds of X called *integral submanifolds*. They are the submanifolds which annihilate all the forms in $\mathscr{E}$. Thus, they satisfy a certain system of differential equations--indeed, this approach is a marvelous geometrization of many of the features of traditional differential equation theory. It seems that the geometric features of many applied areas can be expressed in an optimally geometric way in terms of exterior differential systems.

3. THE FIRST AND SECOND VARIATION IN THE DIFFERENTIAL-FORM FORMALISM

Let X be an ordinary manifold and let Y be an oriented manifold with oriented boundary ∂Y. $\mathscr{M}(Y,X)$ denotes the space of (smooth) mappings

$$\phi: Y \to X.$$

Suppose that:

$$\dim Y = n.$$

Let $\mathscr{D}^n(X)$ be the space of n-th degree differential forms on X.
Let θ be an element of $\mathscr{D}^n(X)$. Define a real-valued function

$$\underline{L}: \mathscr{M}(Y,X) \to R$$

by the following formula:

Geometric Unification

$$\underline{L}(\phi) = \int_Y \phi^*(\theta) \quad . \tag{3.1}$$

Our goal is to compute the derivative of this function $\underline{L}$, as ϕ varies on curves in $\mathcal{M}(Y,X)$. We begin with curves generated by vector fields on X.

Let $V \in \mathcal{V}(X)$ be a vector field on X. Let $s \to \exp(sV)$ be the one-parameter group of diffeomorphisms generated by V. Let s denote that parameter. Apply this group to a $\phi \in \mathcal{M}(Y,X)$.

$$\phi_s = \exp(sV)\phi \quad . \tag{3.2}$$

Then,

$$\begin{aligned}
\frac{d}{ds} \underline{L}(\phi_s) &= \frac{d}{ds} \int \phi_s^*(\theta) \\
&= \frac{d}{ds} \int (\exp(sV)\phi)^*(\theta) \\
&= \frac{d}{ds} \int \phi^* \exp(sV)^*(\theta) \\
&= \int \phi^* \frac{d}{ds} (\exp(sV)^*(\theta)) \\
&= \int \phi^* (\mathscr{L}_V(\exp(sV)^*(\theta))) \\
&= \int \phi^* (i(V)(d(\exp(sV)^*(\theta))) + d(i(V)(\exp(sV)^*(\theta))) \\
&= \int_Y \phi^* (i(V)(\exp(sV)^*(d\theta))) + \int_{\partial Y} \phi^*(i(V)(\exp(sV)^*(\theta)))
\end{aligned} \tag{3.3}$$

(3.3) is the *first variation formula*.

Let us now compute the second variation by differentiating (3.3) again:

$$\frac{d^2}{ds^2} \underline{L}(\phi_s) = \int_Y \phi^*[i(V)(di(V)(\exp(sV)^*(d\theta))] \\
+ \int_{\partial Y} \phi^*[i(V)di(V)(\exp(sV)^*(\theta))] \tag{3.4}$$

Set:

$$\mathbf{f}(V,\theta) = \phi^*(i(V)(\theta)) \tag{3.5}$$

$$\mathscr{S}(V_1,V_2,\theta) = \frac{1}{2}\phi^*(i(V_1)di(V_2)(\theta) + i(V_2)di(V_1)(\theta)) \tag{3.6}$$

For fixed differential form θ and map ϕ, $\mathbf{f}$ is a linear, $\mathscr{S}$ a bilinear differential operator mapping vector fields on X to differential forms on Y. They are the basic geometric objects of the calculus of variations. We call them the *first* and *second variation* differential operators.

<u>Theorem 3.1</u>. (First variation formula):

$$\frac{d}{ds}L(\phi_s) = \int_Y \mathbf{f}(V,d\theta) + \int_{\partial Y}\phi^*(i(V)\theta) \quad , \quad s = 0 \tag{3.7}$$

<u>Theorem 3.2</u>. (Second variation formula):

$$\frac{d^2}{ds^2}L(\phi_s) = \int_Y \mathscr{S}(V,V,d\theta) + \int_{\partial Y}\mathscr{S}(V,V,\theta) \quad , \quad s = 0 \tag{3.8}$$

<u>Remark</u>. Notice that the mappings $\mathbf{f}$ and $\mathscr{S}$ only depend on the values the vector field V takes on the subset $\phi(Y)$.

4. THE FIRST AND SECOND VARIATION FOR "CANONICAL" VARIATIONAL PROBLEMS

Let us now return to the general situation. Let X be a manifold, Y an m-dimensional oriented manifold with boundary ∂Y. Let $\mathscr{E}$ be an exterior differential system on X. Let

$$\mathscr{M}(Y,X,\mathscr{E})$$

be the set of maps

$$\phi: Y \to X \quad ,$$

which are "integral" maps of $\mathscr{E}$, i.e., which satisfy

$$\phi^*(\mathscr{E}) = 0 \quad . \tag{4.1}$$

Let θ be an m-differential form on X. Define a real-valued function

$$\underline{\theta}: \mathscr{M}(Y,X,\mathscr{E}) \to R$$

by integration:

$$\underline{\theta}(\phi) = \int_Y \phi^*(\theta) \quad . \tag{4.2}$$

Geometric Unification

Our goal is to define an appropriate subset of elements of $\mathcal{M}(Y,X,\mathcal{E})$ which can serve as "extremals" of the calculus of variations problem of "extremizing" $\underline{\theta}$, subject to the constraints defined by $\mathcal{E}$.

Of course, the form θ chosen to perform this extremization is not unique; given one, others can be obtained by adding on an arbitrary form in $\mathcal{E}$. Here is a concept which is useful in certain situations in reducing this indeterminancy in the choice of θ.

<u>Definition</u>. Let $\mathcal{I}$ be a Grassmann ideal (but not a differential ideal) of differential forms on X, which is contained in $\mathcal{E}$. The m-form θ is said to be *canonical* (with respect to $\mathcal{I}$) if the following condition is satisfied:

$$d\theta \in \mathcal{I} \quad . \tag{4.3}$$

Let us investigate the meaning of condition (4.3) for the calculus of variations. Suppose that

$$\theta_1, \ldots, \theta_n$$

are a set of forms in $\mathcal{I}$ which generate it as an ideal in $\mathcal{D}(X)$. Then, condition (4.3) means that there are forms

$$\omega_1, \ldots, \omega_n$$

such that:

$$d\theta = \omega_1 \theta_1 + \cdots + \omega_n \theta_n \quad . \tag{4.4}$$

In order to see the significance of condition (4.4) for the extremals, let

$$\phi \in \mathcal{M}(Y,X,\mathcal{E}) \quad .$$

Let $V \in \mathcal{V}(X)$ be a vector field on X, and let

$$s \to \exp(sV)$$

be the one-parameter group of diffeomorphisms of X generated by V. Set

$$\phi_s = \exp(sV)$$

Then, using formula (3.7):

$$\frac{d}{ds} L(\phi_s)\Big|_{s=0} = \int_Y \phi^*(V(\theta)) \tag{4.5}$$

$$= \int_Y \phi^*(i(V)(d\theta)) + \int_{\partial Y} \phi^*(i(V)\theta)$$

Using (4.4),

$$\int_Y \phi^*(i(V)(d\theta)) = \int_Y \phi^*(i(V)(\theta_1\omega_1 + \cdots + \theta_n\omega_n))$$

$$= \int_Y \phi^*(i(V)(\theta_1)\omega_1 + \cdots + i(V)(\theta_n)\omega_n$$
$$\pm \theta_1 i(V)(\omega_1) + \cdots + \theta_n i(V)(\omega_n))$$

$$= \int_Y \phi^*(i(V)(\theta_1)\omega_1 + \cdots + i(V)(\theta_n)\omega_n) \qquad (4.6)$$

since $\phi \in \mathcal{M}(Y, X, \mathcal{E})$, i.e.,

$$\phi^*(\theta_1) = 0 = \cdots = \phi^*(\theta_n) \quad .$$

Notice now that we can make the right hand side of (4.5) vanish by imposing the further condition

$$\phi^*(\omega_1) = 0 = \cdots = \phi^*(\omega_n) \qquad (4.7)$$

on ϕ. This suggests a strategy for defining "extremals" in a rather broad context.

<u>Definition</u>. Let $(\theta, \mathcal{E})$ define a variational problem, as described above. Let $\mathcal{J}$ be a Grassmann ideal of forms of X such that:

$$\mathcal{J} \subset \mathcal{E} \qquad (4.8)$$

> θ is canonical with respect to $\mathcal{J}$, i.e., (4.4) is satisfied.

Let $\mathcal{E}'$ be the exterior differential system defined by $\mathcal{E}$ plus the forms $\omega_1, \ldots, \omega_n$. Then, the map $\phi \in \mathcal{M}(Y, X, \mathcal{E})$ is said to be an *extremal* of the variational problem if:

$$\phi \in \mathcal{M}(Y, X, \mathcal{E}) \quad , \qquad (4.9)$$

i.e., ϕ is an integral map for the exterior differential system $\mathcal{E}'$.

Geometric Unification   153

5. THE SECOND VARIATION FOR VARIATIONAL PROBLEMS IN CANONICAL FORM

Continue with the notation of Section 4. Suppose that:

$$\phi_s = \exp(sV)(\phi)$$

$$d\theta = \theta_1 \omega_1 + \cdots + \theta_n \omega_n$$

with

$$\theta_1, \ldots, \theta_n \in \mathcal{E}$$

$$\phi^*(\omega_i) = 0 \tag{5.1}$$

$$= \phi^*(\theta_i)$$

for $1 \leq i, j \leq n$.

Our goal is to calculate

$$\frac{d^2}{ds^2} L(\phi_s) \Big|_{s=0} \tag{5.2}$$

the *second variation*.

Now,

$$\frac{d}{ds} L(\phi_s) = \frac{d}{ds} \int_Y \phi^*(\exp(sV)^*(\theta))$$

$$= \int_Y \phi^*(\exp(sV)^*(V(\theta)))$$

$$= \int_Y \phi^*(\exp(sV)^*(i(V)(d\theta))) + \int_{\partial Y} \phi^*(\exp(sV)^*(i(V)\theta))$$

Then,

$$\frac{d^2}{ds^2} L(\phi_s) \Big|_{s=0} = \int_Y \phi^*(V(i(V)(d\theta))) + \int_{\partial Y} \phi^*(V(i(V)(\theta))) \tag{5.3}$$

Now, let the vector field V satisfy the following condition:

$$V(\mathcal{E}) \subset \mathcal{E} \tag{5.4}$$

(Geometrically, this means that V is an *infinitesimal symmetry* of the

154   *Geometric Unification*

exterior differential system $\mathscr{E}$ which serves as the constraint set.)

With (5.4) satisfied, and supposing that ϕ is an extremal, in the sense that (4.8) is satisfied, we can give a useful, more explicit form of the first term on the right hand side of (5.3):

$$i(V)(d\theta) = i(V)(\theta_1)\omega_1 \pm \theta_1 i(V)(\omega_1) + \cdots + i(V)(\theta_n)\omega_n \pm \theta_n i(V)(\omega_n)$$

$$\phi^*(V(i(V)(d\theta))) = \phi^*(i(V)(\theta_1)V(\omega_1)) + \cdots + i(V)(\theta_n)V(\omega_n)) \quad (5.5)$$

This is a remarkably simple formula for the Second Variation. Let us sum up as follows:

<u>Theorem 5.1</u>. Let $\phi: Y \to X$ be a map which is an integral map for the exterior differential system spanned by the forms $\theta_1, \omega_1, \ldots, \theta_n, \omega_n$. Let θ be a form such that:

$$d\theta = \theta_1\omega_1 + \cdots + \theta_n\omega_n$$

Let V be a vector on X such that:

$$\phi^*(V(\mathscr{E})) = 0, \quad (5.6)$$

where $\mathscr{E}$ is the exterior differential system generated by the $\theta_1, \ldots, \theta_n$. Let $s \to \phi_s$ be the one-parameter family of maps defined by letting the group generated by V act on ϕ. Then,

$$\frac{d}{ds} \int_Y \phi_s^*(\theta)\Big|_{s=0} = \int_{\partial Y} \phi^*(i(V)\theta) \quad (5.7)$$

$$\frac{d^2}{ds^2} \int_Y \phi_s^*(\theta)\Big|_{s=0} = \int_Y \phi^*(i(V)(\theta_1)V(\omega_1) + \cdots + i(V)(\theta_n)V(\omega_n))$$
$$\quad (5.8)$$
$$+ \int_{\partial Y} \phi^*(V(i(V)\theta))$$

This completes our discussion of the first and second variation formula in the context of the general calculus of variations. We now turn to the general version of another major topic in the classical calculus of variations, the notion of *extremal field*.

6. EXTREMAL FIELDS IN THE DIFFERENTIAL-FORM VARIATIONAL FORMALISM

Suppose that Y, X, θ are as in previous sections and that $\theta_1, \omega_1, \ldots, \theta_n, \omega_n$ are differential forms on X such that:

$$d\theta = \theta_1 \omega_1 + \cdots + \theta_n \omega_n .$$

We have defined and studied single "extremals" as maps $\phi: Y \to X$ such that:

$$\phi^*(\omega_1) = 0 = \cdots = \phi^*(\omega_n) .$$

In the classical calculus of variations, "extremal fields" are *families* of extremals which are tied together with a sort of "integrability" condition. We can now study these families in a general way.

Let Z be another manifold and consider a map

$$\psi: Z \times Y \to X$$

It will be called an *extremal field* if the following conditions is satisfied:

$$\psi^*(d\theta) = 0 . \tag{6.1}$$

The form $\psi^*(\theta)$ is then a closed differential form. We shall suppose that this form is exact, i.e., that there is a form η on $Z \times Y$ such that:

$$\psi^*(\theta) = d\eta . \tag{6.2}$$

Remark. Various interesting possibilities arise following condition (6.1), but not the existence of η *globally*. As explained briefly in [34a], this possibility is linked to a possible "Bohr-Sommerfeld" quantization theory for nonlinear field theories. However, we will not pursue these possibilities here.

Let us now pursue the meaning of (6.2) for the calculus of variations. For each $z \in Z$, let ϕ_z be the map: $y \to \psi(z,y)$ of $Y \to X$. Then,

$$\int_Y \phi_z^*(\theta) = , \text{ using Stokes' formula and relation (6.2),}$$

$$\int_{\phi_z(\partial Y)} \eta \tag{6.3}$$

(The right hand side of (6.3) of course means the integral of the form η over the subset $\phi_z(\partial Y)$.)

This is the very important relation that is the heart of the "extremal field" concept. It says that the function

$$z \to \int_{\phi_z^*} \theta$$

which physically is the "action" of the field represented by θ along the submanifold ϕ_z, *depends only on the values that* η *takes on the boundary of the submanifold.*

In the simplest case, the calculus of variations problems of particle mechanics or "lumped parameter" engineering system, Y is one-dimensional, and η is a zero-form, i.e., a function. It is, typically, a solution of the Hamilton-Jacobi equation. In control theory, one says that this function is the payoff, or cost function. Constructing a synthesis or feedback control of the system amounts to solving the partial differential equation which this function satisfies.

We now leave this general framework in order to pursue certain relations with analytical mechanics.

7. NEWTON'S LAWS AND CHARACTERISTICS OF TWO-DIFFERENTIAL FORMS

In Whittaker's treatise, *Analytical Mechanics*, one finds an admirably general version of the dynamical laws of mechanics, in the Lagrangian framework. In the recent literature, it is more customary to work in this Hamiltonian picture, which excludes certain systems. In fact, Whittaker's approach can be made completely consistent with modern differential geometry. In this section I will show how to do this. (This material is a slightly improved version of that in *Geometry, Physics and Systems*.)

For simplicity, work with systems with one configuration coordinate q and velocity coordinate $\dot{q}$. (The formulas can be readily extended to the general case using vectorial notation.) Consider given two functions $L(q,\dot{q},t)$ and $F(q,\dot{q},t)$ of the variables $(q,\dot{q},t)$. (As explained in the next section, the correct geometric interpretation of these variables is that they are the natural coordinates on a suitable space of one-jets.) The dynamic equations are then:

$$\frac{d}{dt}(\partial_{\dot{q}}(L)) - \partial_q(L) = F \qquad (7.1)$$

Set:

$$\omega = (d(\partial_{\dot{q}}(L)) - \partial_q(L)dt - Fdt)(dq - \dot{q}dt) \ .$$

The solution curves of (7.1) are then the curves $t \to x(t)$ whose tangent vector curves are characteristic for ω, i.e., which satisfy:

$$i(dx/dt)(\omega) = 0 \ .$$

In addition, these curves annihilate the contact form $dq - \dot{q}dt$. We now go into some ways of formulating these ideas in a general geometric manner.

8. THE ONE-JET SPACE AND ITS CONTACT FORMS

Let Q be a manifold, and let (q^i), $1 \leq i,j \leq n$, be a coordinate system of functions on Q. We shall construct a "prolonged" coordinate system on $J^1(R,Q)$, the space of one-jets of mappings: $R \to Q$.

Parameterize R, the "source" space, by the parameter τ. Denote an element of $M(R,Q)$, the space of mappings $R \to Q$, by $q: \tau \to q(\tau)$. Recall that $J^1(R,Q)$ is the quotient of $R \times M(R,Q)$ by the following equivalence relation, namely:

$(t,\mathbf{q})$ is equivalent to $(t',\mathbf{q'})$ if and only if $t = t'$, and $\mathbf{q}, \mathbf{q'}$ have the same tangent vector as t.

Define real-valued functions

$$q^i, \dot{q}^i, t: R \times M(R,Q) \to R$$

by the following formulas:

$$q^i(\tau,\mathbf{q}) = q^i(q(\tau)) \ ,$$

$$\dot{q}^i(\tau,\mathbf{q}) = \frac{d}{d\tau}(q^i(\mathbf{q}(\tau)))$$

$$t(\tau,\mathbf{q}) = \tau \ .$$

Notice that these functions pass to the quotient to define functions on $J^1(R,Q)$; We shall use the same notation for these functions. One now proves readily that:

The functions $(q^i, \dot{q}^i, t)$ define a coordinate system for $J^1(R,Q)$.

In terms of this natural "prolonged" coordinate system for $J^1(R,Q)$, let us now define the following differential forms:

$$\rho^i = dq^i - \dot{q}^i dt \ .$$

Define $\mathscr{C}$ as the Grassmann algebra ideal of differential forms generated by the one-differential forms ρ^i. (Thus $\mathscr{C}$ is not a "differential" ideal, i.e., $\mathscr{C}$ is not closed under the exterior differentiation operation.) Let $\mathscr{C}^k$ denote the subset of $\mathscr{C}$ consisting of those differential forms of degree k. It is readily seen that $\mathscr{C}$ is independent of the coordinate system used to define it.

9. THE STATEMENT OF NEWTON'S LAWS IN TERMS OF THE ONE-JET SPACE

With the differential geometric notion of "contact ideal" on the "one-jet space", we are prepared to state Newton's Laws in what may well be a definitive form. Let Q be a manifold, and let R be the real numbers, thought of as parameterizing physical time. Let $J^1(R,Q) = X$ be the space of one-jets of mappings from R to Q. Let $\mathscr{C}(Q)$ be the Grassmann algebra ideal of differential forms on $J^1(R,Q)$ generated by the contact differential forms, and let $\mathscr{C}^k(Q)$ denote those forms which have degree k.

Definition. A *Newtonian system* with Q as *configuration manifold* is defined as an element ω of $\mathscr{C}^2(Q)$. A curve $\mathbf{x}: \tau \to x(\tau)$ in X is said to be a *trajectory* of the system if the following conditions are satisfied:

$\mathbf{x}$ is an integral submanifold of , i.e.,

$$\rho\left(\frac{d}{d\tau}(\mathbf{x})\right) = 0$$

for $\rho \in \mathscr{C}^1(Q)$.

$$i\left(\frac{d}{d\tau}(\mathbf{x})\right)(\omega) = 0 \quad,$$

i.e., the tangent vector to the curve $\mathbf{x}$ is "algebraically" characteristic for the two-differential form ω.

Remark. Keep in mind that the condition that ω be closed is not imposed, so that one must make a distinction between "algebraic" characteristics for the two-differential form ω and Cauchy characteristics for the exterior differential system generated by the differential form ω.

Instead of working with individual curves satisfying the conditions described above, we shall work with vector fields whose orbit curves satisfy these conditions. Thus, we are looking for vector fields V on $J^1(R,Q)$ which satisfy the following conditions:

$$\mathscr{C}(Q)(V) = 0 \qquad (9.1)$$

$$\mathscr{V}(t) = 1 \qquad (9.2)$$

$$i(V)(\omega) = 0 \quad . \qquad (9.3)$$

Geometric Unification   159

Definition. The vector fields V satisfying conditions (9.1)-(9.3) are called the *Newton-Lagrange* vector fields associated with the differential form ω and the contact ideal $\mathscr{C}(Q)$. The set of all such vector fields is denoted as $NL(\omega)$.

10. MECHANICAL SYSTEMS WHICH HAVE THE SAME NEWTON-LAGRANGE VECTOR FIELD

In the previous section we have assigned to each two-differential form ω which lies in the Grassmann algebra ideal generated by the contact forms on $J^1(R,Q)$ a subset $NL(\omega)$ of the set $V(J^1(R,Q))$. For the purposes of mechanics, it is these vector fields which are the primary objects; the differential form ω is, in a sense, a superfluous object. However, it--or some equivalent structure--seems essential to capturing the differential geometric essence of mechanics. Thus, a natural question suggests itself: How may two such differential forms ω and ω' give rise to the same "dynamics"? Mathematically, this means that we set up an equivalence relation in $\mathscr{C}^2(Q)$: Two elements ω and ω' are equivalent if and only if:

$$NL(\omega) = NL(\omega') \quad . \tag{10.1}$$

Of course, one way of assuring condition (10.1) is to require that the following condition be satisfied:

$$\omega - \omega' \text{ lies in the Grassmann subalgebra generated by } \mathscr{C}(J^1(R,Q)). \tag{10.2}$$

Suppose that Q is an n-dimensional manifold. (Physicists say then that the mechanical system has n degrees of freedom.) $X = J^1(R,Q)$ is then of dimension $2n+1$. The maximal rank that a two-differential form on X could be is $2n$. The space of characteristic vectors is then one-dimensional. In this case, let us say that a mechanics system defined by a two-differential form ω is *regular* if ω has rank $2n$.

Let ω be regular in this sense. Let ω_1, ω_2 be differential forms which do not involve dt such that

$$\omega = \omega_1 + \omega_2 \, dt \quad .$$

Thus, a vector field V belongs to $NL(\omega)$ if and only if:

$$i(V)(\omega_1) = \omega_2 \quad . \tag{10.3}$$

Theorem 10.1. ω is non-singular if and only if the two-differential form ω_1 is of rank $2n$.

Proof. Linear algebra.

Here is another more explicit way of determining all equivalent two-forms for a non-singular mechanical system. Again, suppose that ω is given and that it is regular, so that the NL vector field V is unique. As one of its defining properties, we know that V annihilates the contact ideal. Suppose that θ_j, $1 \leq j,k \leq n$, is a basis of one-differential forms in the contact ideal. We can then (algebraically) find n more one-differential forms ξ^j such that:

$$\omega^j(V) = 0$$

and

$$(\theta^j, \omega^j, dt)$$

form a basis for one-dimensional forms.

<u>Theorem 10.2</u>. Any two-differential form ω' which lies in $\mathscr{C}(Q)$ and is equivalent to ω is of the following form:

$$\omega' = a_{jk} \theta^j \theta^k + b_{jk} \omega^j \omega^k \quad ,$$

where the coefficients a and b are arbitrary matrices of functions.

Proof. Again, this is trivial linear algebra.

Remark. We see from this that the orbit curves of V, i.e., the trajectories of the mechanics system, form a one-dimensional foliation of J. It is defined by the following Pfaffian equations:

$$\theta^j = \omega^j = 0 \quad ,$$
$$j = 1, \ldots, n \quad .$$

This is the form that Cartan would write the equations in.

11. SYMPLECTIC STRUCTURES ON THE SPACE OF ALL TRAJECTORIES OF A REGULAR MECHANICAL SYSTEM

Continue with Q as an n-dimensional manifold which is the configuration space of a mechanics system. Let $X = J^1(R,Q)$. Let ω be a two-differential form of rank $2n$ which lies in $\mathscr{C}(Q)$, hence which defines a regular mechanical system. The NL vector field V is then uniquely determined.

The vector field V defines a foliation of X, i.e., the leaves of the foliation are one-dimensional, and are the orbit curves of V. We shall suppose that the foliation is <i>regular</i> in the sense that the quotient space of the foliation exists as a Hausforff manifold; denote it by X'.

Geometric Unification

As we shall see, it will be important for physical purposes to know if there is an equivalent two-differential form ω' to ω which is closed, i.e., satisfies the following condition:

$$d\omega' = 0 \ .$$

Such a two-differential form will then satisfy the following further relation:

$$V(\omega') = i(V)(d\omega') + d(i(V)(\omega'))$$
$$= 0 \ ,$$

i.e., ω' is the pull-back of a differential form on X', i.e., ω' lives on X'. Although this is a trivial observation, it is worth stating in the following form:

<u>Theorem 11.1</u>. Any closed two-differential form ω' equivalent to ω defines a (possibly singular) symplectic structure on the space X' of all trajectories of the mechanics system.

Conversely, we might start off with a closed two-differential form ω' on X' and pull it back to X via the quotient map: $X \to X'$. The following condition will be satisfied automatically:

$$i(V)(\omega') = 0 \ .$$

However, we will not know whether it is equivalent to ω. For this to happen ω' must, in addition, belong to the contact ideal. For this to happen, there must be n one-differential forms ξ'^j, $j,k = 1,\ldots,n$, such that the following condition is satisfied:

$$\omega' = \xi'^j \rho^j \ ,$$

and

$$\xi'^j(V) = 0 \ .$$

In any case, we see that it is an interesting and important question to decide when there is another two-differential form ω' equivalent to the one given to us via the mechanical system and which is closed. In fact, we shall see that this is a general form of the question often asked by physicists: When is a given mechanical system "conservative" in the sense that its equations are of the Lagrangian form? As we shall see, the existence of a ω' with the properties described above is a necessary condition. We shall return to the formalism of Whittaker's treatise to investigate this question in one direction.

12. WHEN IS THE TWO-DIFFERENTIAL FORM DEFINED BY THE NEWTON-LAGRANGE EQUATIONS CLOSED?

Keep the notation of previous sections, but work with local coordinates $(q, \dot{q}, t)$ for $J^1(R, Q)$. The contact differential forms are then:

$$\rho^j = dq^i - \dot{q}^j \, dt \quad .$$

Let V be a vector field such that $V(t) = 1$, and such that the orbits of V are the solutions of the Newton-Lagrange equations:

$$\frac{d}{dt}(\partial_{\dot{q}} L) - \partial_q L = F \quad .$$

We can set:

$$\xi_j = d\left(\partial_{\dot{q}^j}(L)\right) - \partial_{q^j}(L) \, dt - F_j \, dt \quad ,$$

and

$$\omega = \xi_j \rho^j \quad .$$

We already know that:

$$i(V)(\omega) = 0 \quad .$$

We now ask: What are the conditions that ω be a closed differential form?

Now,

$$\omega = \omega_1 + \omega_2 \quad ,$$

with

$$\omega_1 = \xi_j \rho^j$$

and

$$\omega_2 = -F_j \, dt \rho^j \quad .$$

Notice that ω_1 is the contribution of the Lagrangian, i.e., the "kinetic energy", while ω_2 is due to the "forces". (Of course, one could also try to separate contributions of "internal" and "external" forces; typically the former might go into L, while the latter are in F. However, our goal here is to isolate the mathematical issues rather than become extensively involved with the physics.)

We can now see that ω_1 is closed. To this end, set:

$$p_j = \partial_{\dot{q}^j}(L) \quad .$$

Then,

$$\omega_1 = \left(dp_j - \partial_{\dot{q}^j}(L)\,dt\right)(dq^j - \dot{q}^j\,dt)$$

$$= dp_j\,dq^j - \partial_{\dot{q}^j}(L)\,dt\,dq^j - dp_j\,\dot{q}^j\,dt$$

$$= dp_j\,dq^j - dh\,dt \quad,$$

with

$$h = L - p_j\dot{q}^j \quad.$$

This formula (which also is the key one for Cartan's method of relating the Lagrange and Hamilton equations) shows that:

$$d\omega_1 = 0 \quad,$$

as required. Let us sum up what we have done so far:

Theorem 12.1. From the Lagrangian and the force law F, construct the two-differential form ω. It is closed, i.e., defines a symplectic structure on the space of trajectories of the mechanical system, if and only if:

$$d(F_j\,dt\,(dq^j - \dot{q}^j\,dt)) = 0 \quad. \tag{12.1}$$

Let us work out condition (12.1) more explicitly.

$$0 = \left(\partial_{\dot{q}^k}(F_j)\,d\dot{q}^k + \partial_{q^k}(F_j)\,dq^k\right)dt\,(dq^j - \dot{q}^j\,dt) + F_j\,dt\,d\dot{q}^j\,dt$$

$$= \left(\partial_{\dot{q}^k}(F_j)\,d\dot{q}^k + \partial_{q^k}(F_j)\,dq^k\right)dq^j \quad.$$

This relation then forces the following condition:

$$\partial_{\dot{q}^k}(F_j) = 0 \quad,$$

i.e., F_j is a function of the q's alone.

Finally, we come to the following condition:

$$\partial_{q^k}(F_j) = \partial_{q^j}(F_k) \quad.$$

These conditions together say that the one-differential form

$$F_j\, dq^j = F$$

is a closed differential form which lives on Q. Thus, locally there is a function $V(q)$ such that

$$dV = F .$$

V is the familiar "potential" for the forces. Thus, we see that imposing the condition $d\omega = 0$ does indeed, in this case, lead to the "conservative" type of mechanical system, since the Lagrangian L can be modified so that the equations are of the pure L form, with no "forces". (Of course, if Q does not have zero first Betti number, so that V need not exist globally, there are interesting topological ramifications to the theory.)

Now, let us explore possible modifications of the two-differential form ω which will leave unchanged the equations of motion, but will have the possibility of making it closed. Here is one such transformation:

$$\omega' = \omega + a_{jk}\rho^j\rho^k ,$$

where (a) is a skew-symmetric matrix of functions. From what has been proved above, we see that:

$$d\omega' = -d(F_j dt(dq^j - \dot{q}^j dt)) + d(a_{jk}\rho^j\rho^k)$$

$$= dF_j\, dq^j\, dt + da_{jk}\rho^j\rho^k + 2a_{jk}\, dt\, d\dot{q}^j\, dq^k .$$

<u>Theorem 12.2.</u> $d\omega' = 0$ if and only if the following conditions are satisfied:

$$\partial_{q^l}(a_{jk})\, dq^l\, dq^j\, dq^k = 0 \qquad (12.2)$$

$$\partial_{\dot{q}^l}(a_{jk}) = 0 \qquad (12.3)$$

$$\partial_{\dot{q}^k}(F_j) = 2a_{kj} \qquad (12.4)$$

$$\partial_t(a_{lk}) = \frac{1}{2}\left(\partial_{q^l}(F_k - 2a_{jk}\dot{q}^j) - \partial_{q^k}(F_l - 2a_{jl}\dot{q}^j)\right) \qquad (12.5)$$

<u>Proof</u>. These relations are obtained by equating to zero the terms in the relation $d\omega' = 0$ which involve $dq\, dq\, dq$, $d\dot{q}\, dq\, dt$, and $dq\, dq\, dt$.

Geometric Unification

We can now work on relations (12.2)-(12.5). First, relations (12.3) and (12.4) together imply that the force terms are of the following type:

$$F_k = 2a_{jk}\dot{q}^j + b_k , \qquad (12.6)$$

where the a's and b's are functions of q and t alone.

Theorem 12.3. With the force law F of form (12.6), and with condition (12.2) satisfied, the following conditions are necessary and sufficient that the two-differential form ω' be closed:

$$\partial_{q^l}(b_k) - \partial_{q^k}(b_l) = \partial_t(a_{lk}) . \qquad (12.7)$$

Proof. Substitute the Ansatz (12.6) into the other conditions.

An important case physically is that where the force law is time-independent. In this case, the conditions found above simplify and can be written in an elegant "global" form.

Theorem 12.4. Suppose that we define a mechanical system by means of a Lagrangian function L and a force law F, with $\partial_t(F) = 0$. Suppose that this force law satisfies condition (12.6). Define differential forms α and β on Q so that:

$$\alpha = a_{jk} dq^j dq^k ,$$

$$\beta = b_j dq^j .$$

Then, the condition that the two-differential form ω' be closed is that the following conditions be satisfied:

$$d\alpha = d\beta = 0 .$$

If $d\omega' = 0$, then the trajectories of the mechanical system are Cauchy characteristic curves of ω'. Assuming that the trajectories define a regular foliation of $J^1(R,Q)$, the quotient space of the foliation has a symplectic structure.

The conditions we have found for the existence of a symplectic structure on the space of trajectories is especially important in electromagnetic theory. For example, suppose that q are the coordinates of a charged particle in R^3. It is well-known that the force law then has the form described in condition (12.6), where the a's are the components of the "magnetic" field, the b's the "electric" field. The conditions $0 = d\alpha = d\beta$ are then part of

Maxwell's equations. It is well known that to write the charged particle equations in Hamiltonian form requires that these differential forms be exact; the forms whose exterior derivatives gives them are the potentials. Thus, we see that the charged particle equations themselves may be given a symplectic structure in a "global" way without the intervention of the potentials. Such a possibility is very relevant to the theory of quantization, particularly the "Bohm-Aharonov effect".

13. THE OPTIMAL CONTROL PROBLEM IN CANONICAL FORM

The usual form of the optimal control problem may be described as follows: Extremize

$$\int_0^T L(x,u,t)\, dt + F(x(T)) \tag{13.1}$$

subject to the input-constraint

$$\frac{dx}{dt} = f(x,u,t)\quad , \tag{13.2}$$

$$x \in R^n, \quad u \in R^n, \quad t \in R \quad .$$

This is, of course, a classical calculus of variations problem (of "Bolza" type), and can be treated in many ways. As another illustration of the general formalism I now want to discuss how this can be written in terms of differential forms. (See the excellent book *Singular Optimal Control Problems* by Bell and Jacobson [4] for the traditional material.)

Let X be R^{n+m+1} be the space of variables (x,y,u,t). Put indices on the variables

$$x = (x^i) \quad , \qquad 1 \le i, j \le n$$

$$u = (u^a) \quad , \qquad 1 \le a, b \le m \quad .$$

$$y = (y_i) \quad .$$

We can then write (13.2) as a differential form constraint:

$$\theta^i \equiv dx^i - f^i(x,u)\, dt = 0 \quad . \tag{13.3}$$

Set:

$$\theta = L\, dt + y_i \theta^i \tag{13.4}$$

Geometric Unification

If

$$t \to (x(t), y(t), u(t), t) = \phi(t)$$

is a curve in X, the quantity (13.1) to be extremized is then

$$\int_0^T \phi^*(\theta) + F(x(T)) \qquad (13.5)$$

where $x \to F(x)$ is a real-valued function. In order to apply the general formalism developed in this paper, we now want to write θ in "canonical form", i.e., write it as follows:

$$\theta = \omega_i \theta^i , \qquad (13.6)$$

where ω^i are one-forms.

Now,

$$d(L\,dt) = \left(L_{x^i} dx^i + L_{u^a} du^a \right) dt \qquad (13.7)$$

(Subscripts denote partial derivatives.) Thus,

$$\begin{aligned}
d\theta &= \left(L_{x^i} dx^i + L_{u^a} du^a \right) dt + dy_i(\theta^i) - y_i \left(f^i_{x^j} dx^j dt + f^i_{u^a} du^a dt \right) \\
&= \left(L_{x^i} \theta^i + L_{u^a} du^a \right) dt + dy_i \theta^i - y_i f^i_{x^j} \theta^j dt - y_i f^i_{u^a} du^a dt \\
&= \theta^i \left(L_{x^i} dt + L_{u^a} du^a - dy_i - y_j f^j_{x^i} dt \right) + \left(L_{u^a} - y_i f^i_{u^a} \right) du^a \, dt
\end{aligned} \qquad (13.8)$$

In order to put this in canonical form, let X' be the subset of X determined by the following relations:

$$L_{u^a} - y_i f^i_{u^a} = 0 . \qquad (13.9)$$

Assume that X' is a submanifold of X. Restrict all differential forms to X', but, for notational simplicity, denote them by the same symbol. We then have the following result:

<u>Theorem 13.1.</u> With the above assumptions,

$$d\theta = \theta^i \omega_i , \qquad (13.10)$$

with

$$\omega_i = L_{x^i} dt + L_{u^a} du^a - dy_i + y_j f^j_{x^i} dt$$

$$= \text{, using (13.9)}$$

$$L_{x^i} dt + y_j \left(L_{u^a} du^a + f^j_{x^i} dt \right) - dy_i \qquad (13.11)$$

Let $\mathcal{E}$ be the exterior differential system generated by the forms θ^i, and let $\mathcal{E}'$ be the exterior system generated by $\mathcal{E}$ plus the $\omega_1 \ldots$. Let us define the *extremals* as the integral curves of $\mathcal{E}'$ which are parameterized by t. Let

$$\phi: [0, T] \to X$$

be such an extremal curve. Let us compute the first variation of the functional (13.1). Let V be a vector field,

$$s \to \phi_s = \exp(sV)(\phi)$$

the one-parameter family of curves is generates. Then,

$$\frac{d}{ds} \int \phi_s^*(\theta) \Big|_{s=0} + \frac{d}{ds} F(\phi_s)(T) \Big|_{s=0} = \text{, using the fact that } \phi \text{ is an integral curve of } \mathcal{E}',$$

$$\theta(V)(\phi(T)) - \theta(V)(\phi(0)) \qquad (13.12)$$
$$+ V(F)(\phi(T))$$

Thus, we see that the functional

$$\phi \to \int_0^T \phi^*(\theta) + F(\phi(T))$$

will be extremized relative to variations of curves which satisfy the following boundary conditions:

$$\theta(V)(\phi(0)) = 0$$
$$\theta(V)(\phi(T)) + F(\phi(T)) = 0 \qquad (13.13)$$

For example, let us suppose that θ and V are of the following form:

$$\theta = y_i dx^i - H dt$$
$$V = v^i \frac{\partial}{\partial x^i} + v_i \frac{\partial}{\partial y^i} \qquad (13.14)$$

Geometric Unification

Then, condition (13.13) implies the following one:

$$y_i v^i(\phi(0)) = 0$$

$$y_i v^i(\phi(T)) + \frac{\partial F}{\partial x^i}(\phi(T)) v^i(\phi(T)) = 0 \ .$$
(13.15)

These conditions are implied by the following ones:

$$V(x^i)(\phi(0)) \equiv v^i(\phi(0)) = 0 \tag{13.16}$$

$$y_i(\phi(T)) = -\frac{\partial F}{\partial x^i}(\phi(T)) \tag{13.17}$$

(13.16) means, geometrically, that the variations leave invariant the $t = 0$ value of the state variable x. (13.17) is a final-time boundary condition, requiring prescribing the value of the costate vector.

The extremal equations take the simplest form providing that the variational problem is *non-singular*, in the sense that the matrix

$$(a, b) \rightarrow \left(L_{u^a u^b} - y_i f^i_{u^a u^b} \right)$$

is non-singular at each point of the subset X' determined by relations (13.9). In this case, the implicit function theorem guarantees that X is a submanifold of X, and one easily sees that the one-forms

$$\theta^i, \omega_i$$

are linearly independent. Hence, the two-form $d\theta$ has rank n, and, in suitable local coordinates, the extremal curves equations, i.e., the equations for the Cauchy characteristic curves of $d\theta$, take a Hamiltonian form.

Let us examine what happens in the simplest singular case; namely, suppose that the optimal control problem is quadratic and the control constraints are linear, i.e.,

$$L(x, u) = \frac{1}{2}(Q_{ab} u^a u^b + R_{ij} x^i x^j)$$

$$f^i(x, u) = A^i_j x^j + B^i_a u^a$$
(13.18)

Equation (13.9) takes the following form:

$$Q_{ab}u^b + y_i B^i_a = 0 \tag{13.19}$$

Equations (13.15) are linear, hence automatically define an X' as a submanifold.

$$\omega_i = R_{ij}x^j dt + y_j(Q_{ab}du^b + A^j_i dt) - dy_i \tag{13.20}$$

$$\theta^i = dx^i - (A^i_j x^j + B^i_a u^a) dt$$

Thus, the vector field V of the form

$$V = v^a \frac{\partial}{\partial u^a} ,$$

with

$$Q_{ab} v^b = 0$$

satisfy the following condition:

$$\omega_i(V) = 0 = i(V)(d\omega_i)$$

$$i(V)(d\theta) = 0 .$$

They are then Cauchy characteristic for $d\theta$. The quotient of X' by the foliation determined by these vector fields is a manifold X", and the form $d\theta$ (but not θ!) is the pull-back of a form on X". The extremal curves, when projected down to X", takes their "canonical form".

There is a prime example of such a singular linear problem given to us by nature--the electromagnetic field. Of course, in this case, the state and control space is infinite dimensional. Let us consider it from this point of view.

14. MAXWELL'S EQUATIONS

Let us consider Maxwell's equations for the electromagnetic field in a vacuum. They can be stated in the very elegant form:

$$\delta d\theta = 0 \tag{14.1}$$

where $\theta \in \mathcal{D}^1(R^4)$ is the "electromagnetic potential", and where

$$\delta: \mathcal{D}^2(R^4) \to \mathcal{D}^1(R^4)$$

is the Hodge operator (dual to d) with respect to the Lorentz metric on R^4. It is known that these equations are the extremal equations of the following variational problem: Extremize

$$\int_{R^4} <d\theta, d\theta> \, dx \qquad (14.2)$$

where $<\,,\,>$ is the inner product on differential forms generated by the Lorentz metric, and where

$$dx \in \mathcal{D}^4(R^4)$$

is the volume element differential form defined by the Lorentz metric.

Let R^4 be space-time with coordinates

$$(t, x^i) \, , \qquad 1 \leq i, j \leq 3 \quad .$$

The *Lorentz metric* is the following Riemannian metric.

$$ds^2 = c^2(dt^2) - (dx^1)^2 - (dx^2)^2 - (dx^3)^2 \qquad (14.3)$$

(c is the velocity of light.) This metric defines an inner product $<\,,\,>$ on vector fields and differential forms:

$$<dt, dt> = \frac{1}{c^2}$$

$$<dt, dx^i> = 0 \qquad (14.4)$$

$$<dx^i, dx^j> = \delta^{ij} \quad .$$

Let

$$\theta$$

be a one-form on R^4. We can then write:

$$\theta = \theta_1 + \theta_0 \, dt \qquad (14.5)$$

where θ_1 involves the forms dx^i alone. (Physically, θ_1 is the *magnetic potential*, θ_0 the *electropotential*.)

We can consider θ_1 and θ_0 as forms on R^3, which depend on t as a *parameter*, i.e., as curves with t as parameter in the space of differential forms on R^3. Let

$$\theta_{1,t} \, , \qquad \theta_{0,t}$$

be the derivatives of these forms on R^3 with respect to the parameter t.

Then,

$$d\theta = d_x(\theta_1) + (-\theta_1, + d_x\theta_0)dt \quad (14.6)$$

(Here d_x denotes the exterior derivatives with respect to the variables x^i.)

$$<d\theta,d\theta> = <d_x\theta_1,d_x\theta_1> + c^{-2}<\theta_{1,t} + d_x\theta_0, \theta_{1,t} + d_x\theta_0> \quad (14.7)$$

Now, the Euclidean volume element dx of R^4 splits as follows

$$dx \quad dx^1 dx^2 dx^3 dt$$

$$= d^3x\, dt \quad ,$$

with

$$d_1^3 x = dx^1 dx^2 dx^3$$

Hence, the action for the free electromagnetic field can be written as

$$<d\theta,d\theta>dx = [(-<d_x\theta_1,d_x\theta_1> + c^{-2}<-\theta_{1,t} + d_x\theta_0>)d^3x]\, dt \quad (14.8)$$

(The minus sign is chosen on the right hand side of (14.8) so that the inner product is positive.)

We can abstract from this formula a general variational problem. Let $\Gamma_0, \Gamma_0, \Gamma_2$ be real vector spaces with positive-definite symmetric inner products that we denote as $<\,,\,>$. Set:

$$\Gamma = \Gamma_0 \oplus \Gamma_1 \quad .$$

Given a curve $\underline{\gamma}: t \to \gamma(t) = (\gamma_0(t), \gamma_1(t))$ in Γ, define:

$$\underline{L}(\underline{\gamma}) = \int_a^b \left[-<\alpha\gamma_1, \alpha\gamma_1> + c^{-2} \left\| \frac{d\gamma_1}{d\tau} + \alpha\gamma_0 \right\|^2 \right] dt$$

where α is a linear map of

$$\Gamma_i \to \Gamma_{i+1} \quad , \qquad i = 0,1 \quad .$$

Notice that this has the form of a "linear mechanical system with singular kinetic energy term". (γ_0, γ_1) are the configuration space variables. We can treat this as an "ordinary" calculus of variations problem. The extremals are the solutions of the following differential equations:

Geometric Unification

$$\frac{d}{dt}\left(L_{\dot{\gamma}_1}\right) - L_{\gamma_1} = 0$$

$$\frac{d}{dt}\left(L_{\dot{\gamma}_0}\right) - L_{\gamma_0} = 0 \qquad (14.9)$$

Since the Lagrangian does not depend on $\dot{\gamma}_0$, these extremal equations take the following form:

$$\frac{d}{dt} c^{-2}\left(-\frac{d\gamma_1}{dt} + \alpha\gamma_0\right) = -2T\alpha\gamma_1$$

$$0 = L_{\gamma_0} = -\frac{d}{dt}(\delta\gamma_1) + T\alpha\gamma_0 \qquad (14.10)$$

where α^T is the dual map (with respect to the form $\langle\,,\,\rangle$) to α.

To write these equations in more familiar form, set

$$E = \left(\alpha\gamma_0 - \frac{d\gamma_1}{dt}\right) c^{-1}$$

$$\equiv \text{electric field}$$

$$B = \alpha\gamma_1 \qquad (14.11)$$

$$= \text{magnetic field}.$$

Then, Equations (14.10) are equivalent to:

$$\frac{d}{dt} E = -2c \, \text{curl}(B)$$

$$\frac{d}{dt} B = \text{curl}(E),$$

which are Maxwell's equations in their usual form.

References

1. Anderson, B.D.O. and Moore, J.B., *Linear Optimal Control*, Prentice-Hall, Englewood Cliffs, N.J., 1971.

2. Arens, R., Differential-geometric elements of analytic dynamics, *J. Math. Anal. Appl.*, 9, 1965-202 (1964).

3. Arnold, V., *Methods of Classical Mechanics*, Springer Graduate Texts on Math. No. 60, Springer-Verlag, New York, 1978.

4. Bell, D.J., and Jacobson, D.H., *Singular Optimal Control Problems*, Academic Press, 1975.

5. Bliss, G.A., *Lectures on the Calculus of Variations*, Prentice-Hall, Englewood Cliffs, N.J., 1971.

6. Bryson, A.E., and Ho, Y.C., *Applied Optimal Control*, Blaisdell, Waltham, MA, 1969.

7. Caratheodory, C., *Calculus of Variations and Partial Differential Equations*, Holden-Day, San Francisco, 1965.

8. Cartan, E., *Leçons sur les Invariants Integraux*, Herman, Paris, 1922.

9. Dedecker, P., *Calcul des Variations, Formes Différentielles et Champs Géodésiques*, Coll. Internat. du C.N.R.S., Strasbourg, 1953.

10. _____, Calcul des variations et topologie algébrique, *Mém. Soc. Roy. Sci. Liege* 19 (1957).

11. _____, On the generalization of symplectic geometry to multiple integrals in the calculus of variations, In *Diff. Geom. Methods in Mathematical Phys.*, Springer Lecture Notes in Mathematics, Vol. 570.

12. de Donder, Th., *Théorie Invariante du Calcul des Variations*, Gauthier-Villars, Paris, 1935.

13. Dewitt, B.S., Dynamical theory in curved spaces, I, *Rev. Mod. Phys.*, 26, 377-397 (1957).

14. Dirac, P.A.M., Generalized Hamiltonian dynamics, *Can. J. Math.*, 2, 129-148 (1950).

15. _____, *Lectures on Quantum Mechanics*, Belfer Graduate School of Sci., Monograph Series Vol. 2, Yeshiva Univ., N.Y., 1964.

16. Droz-Vincent, P., Hamiltonian systems of relativistic particles, *Rep. Math. Phys.*, 8, 79-101, 1975.

17. Duistermaat, J.J., On the Morse index in variational calculus, *Adv. Math.*, 21, 173-195 (1976).

18. Estabrook, F.B., and Wahlquist, H.D., The geometric approach to sets of ordinary differential equations and Hamiltonian dynamics, *SIAM Review*, 17(2), 201-220 (1975).

19. Gabasov, R. and Kirillova, F.M., High order necessary conditions for optimality, *SIAM J. Control*, 10, 127-168 (1972).

20. Garcia, P.L., The Poincare-Cartan invariant in the calculus of variations, *Symp. Math.*, Vol. 14, Instituto Nazion. di Alta Mate. Roma, pp. 219-246, 1974.

21. _____, Gauge algebras, curvature and symplectic structure, *J. Diff. Geom.*, 12, 351-359 (1974).

22. _____, P.L., and Perez-Rendon, A., Symplectic approach to the theory of quantized fields, I, *Comm. Math. Phys.*, 13, 22-44 (1969).

23. _____, Symplectic approach to the theory of quantized fields, II, *Arch. Rat. Mech. Anal.* 43, 101-124 (1969).

24. Gelfand, I.M., and Fomin, S., *Calculus of Variations*, Prentice-Hall, Englewood Cliffs, N.J., 1963.

25. Godbillon, C., *Geometrie Differentielle et Mecanique Analytique*, Herman, Paris, 1969.

26. Goh, B.S., The second variation for the singular Bolza problem, *SIAM J. Control*, 4, 309-325 (1966).

27. _____, Optimal singular control for multi-input linear systems, *J. Math. Analysis Applic.*, 20, 534-539 (1967).

28. Goldschmidt, H. and Sternberg, S., The Hamilton-Jacobi formalism in the calculus of variations, *Ann. Inst. Fourier*, 23, 203-267 (1973).

29. Hermann, R., Some differential geometric aspects of the Lagrange variational problem, *Illinois J. Math.*, 6, 634-673 (1962).

30. _____, The second variation for variational problems in canonical form, *Bull. Amer. Math. Soc.*, 71, 145-148 (1965).

31. _____, The second variation for minimal submanifolds, *J. Math. Mech.*, 16, 473-492 (1966).

32. _____, *Differential Geometry and the Calculus of Variations*, Academic Press, New York, 1969.

33. _____, *Geometry, Physics and Systems*, Marcel Dekker, New York, 1973.

34. _____, Some remarks on the geometry of systems, *Proc. NATO Instit. on Geometric and Algebraic Methods for Nonlinear Systems*, Imperial College, D. Mayne (ed.), Reidel, Boston MA, 1973.

35. Hermes, H., Controllability and the singular problem, *SIAM J. Control*, 1964.

36. Kelley, H.J., Kopp, R.E., and Moyer, H.G., Singular extremals, in *Topics in Optimization*, G. Leitmann (ed.), Academic Press, New York, 1967, pp. 63-101.

37. Krener, A.J., The high order maximal principle, in *Geometric Methods in Systems Theory*, D.Q. Mayne and R.W. Brockett (eds.), NATO Advanced Studies Institute Series: Mathematics and Physics, 1973, pp. 174-184.

38. Kunzle, H., Degenerate Lagrangian systems, *Ann. Inst. H. Poincare (A)*, 11, 393-414 (1969).

39. Lanczos, C., *The variational principles of mechanics*, 2nd ed., Univ. of Toronto Press, 1962.

40. Mayne, D.Q., and R.W. Brockett (eds.), *Geometric Methods in System Theory*, Dordrecht, Holland: Reidel, 1973.

41. Pontryagin, L.S., Boltyanskii, V.G., Gamkrelidze, R.V., and Mishchenko, E.F., *The Mathematical Theory of Optimal Process*, Wiley Interscience, New York, 1962.

42. Souriau, J.M., *Structure des systemes dynamiques*, Dunod, Paris, 1970.

43. Sternberg, S., *Lectures on Differential Geometry*, Prentice-Hall, Englewood Cliffs, N.H., 1964.

44. Valentine, F.Q., The problem of Lagrange with Differential inequalities as added side conditions, in *Contributions to the Theory of Calculus of Variations (1933-1937)*, Univ. Chicago Press, Chicago, 1937, pp. 403-447.

45. Weyl, H., *Ann. of Math.*, 36, 607 (1935).

46. Whittaker, E., *Analytical Dynamics*, Cambridge Univ. Press,

47. Abraham, R. and Marsden, J.E., *Foundations of Mechanics*, 2nd ed., Addison-Wesley, Reading, MA, 1979.

Chapter 11

MAXWELL'S ELECTROMAGNETIC EQUATIONS AND ANALYTICAL MECHANICS

1. INTRODUCTION

Freeman Dyson has remarked that it is one of the "missed opportunities" that mathematicians never took up Maxwell's theory of electromagnetism in a serious way. To this day, there does not exist a development of the mathematical foundations of Maxwell theory which is even remotely comparable to what is available for other branches of physics, e.g., analytical and quantum mechanics. Surely one of the elements that must play a major role in any such ultimate development is the relation to Hodge's theory of "harmonic integrals". I have already (e.g., in *Interdisciplinary Mathematics*, Vols. 4, 5, and 19) presented certain relations between the Hodge and Maxwell theory. In this chapter I want to proceed further, particularly emphasizing the nature of *energy* in the Maxwell-Hodge theory.

Another intriguing feature of the Maxwell theory is the relation between the Maxwell equations and the electrical circuit equations. One is a set of *partial* differential equations, the other *ordinary*. There seem to be various schemes for deriving the latter in terms of the former. The most systematic -- and the only one I can really understand -- proceeds via the variational formulation. This approach was developed by Gabriel Kron and K. Kondo (in the RAAG memoirs). It is most readily found in the book *Non-Holonomic Dynamics*, by Neimark and Fufiev. It seems to me that this method of deriving ordinary differential equations from partial is extremely interesting from the general point of view of differential-geometric foundations of science and technology. (For example, it has much in common with the much touted "finite element method".) The key idea is to take a multiple integral variational problem and, by constraining the functions allowed in the competition, obtain a single integral problem. Thus we seem to derive a set of ordinary differential equations from a system of partials, by a basically *geometric* mechanism.

2. THE LAGRANGIAN FOR MAXWELL THEORY

Let X be a manifold. $\mathcal{D}^r(X)$, $r = 0,1,\ldots$, denotes the r-th degree differential forms. *Exterior multiplication* is denoted by simple juxtaposition, not by a special symbol as in most differential geometry texts. Let $\langle , \rangle$ denote a Riemannian metric on X. It defines an inner product -- also denoted by $\langle , \rangle$ -- on all tensor fields. We will especially use the inner product

$$\theta_1, \theta_2 \to \langle \theta_1, \theta_2 \rangle$$

on differential forms.

Fix an integer m. Set:

$$L(\theta) = \int [\langle d\theta, d\theta \rangle + 2\langle \theta, J \rangle] \, dx \, , \qquad (2.1)$$

where dx is the volume element form of the Riemannian metric and where is an m-form on X. J is a fixed m-form called the *current*. L is the *Lagrangian* of Maxwell theory. We shall "extremize" it, following the style of classical mathematical physics/calculus of variations.

3. FIRST VARIATION OF THE LAGRANGIAN AND MAXWELL'S EQUATIONS

Let α be another m-form on X. Set:

$$L(\theta + \varepsilon\alpha) = L(\theta) + \varepsilon \partial L(\omega, \alpha) + \varepsilon^2 \partial^2 L(\theta, \alpha) \, .$$

∂L and $\partial^2 L$ are the *first* and *second variations*.

From (2.1) we see that:

$$\partial L(\theta, \alpha) = 2 \int (\langle J - \delta d\theta, \alpha \rangle \, . \qquad (3.1)$$

(When no confusion is likely, we shall omit the volume element differential form, i.e., dx, from the formulas. δ is the *Hodge dual* to the exterior derivative. It is the generalized form of the *divergence*.) Formulas (3.1) are the *first variation formulas*. The condition that this vanish for all α, i.e.,

$$d\theta = J \, , \qquad (3.2)$$

is Maxwell's equations. θ is the *electromagnetic potential*, $\omega = d\theta$ the *electromagnetic field*.

4. THE ALGEBRA OF MAXWELL'S EQUATIONS

Formula (3.2) exhibits Maxwell's equations of electromagnetism in maximally elegant form -- from the physicist's point of view it is *too* elegant. The trouble is that X and its Riemannian metric are very general. The manifolds and metrics which appear in physics have a special structure. Part of this structure involves what one might call a "time" structure. In this general framework, this can be interpreted algebraically in the following way.

Maxwell's Equations

Let

$$m = 1 \ .$$

Let τ be a real-valued function on X such that:

$$\langle d\tau, d\tau \rangle = -1 \ . \tag{4.1}$$

(Thus, "τ" must be a solution of the Hamilton-Jacobi equation for the metric, i.e., the orbits of the vector field

$$\text{grad } \tau$$

are *geodesics*.)

Let ω be a two-form. Decompose it algebraically as:

$$\omega = \omega_B + \omega_E \, d\tau \ , \tag{4.2}$$

with:

$$\langle \omega_B, d\tau \rangle = 0$$

$$= \langle \omega_E, d\tau \rangle \tag{4.3}$$

Let α be a compactly supported one-form. α can be decomposed as follows:

$$\alpha = \alpha' + \beta \, d\tau \tag{4.4}$$

with:

$$\langle \alpha', d\tau \rangle = 0 \tag{4.5}$$

$$\beta \in \mathscr{F}(X) \ .$$

Then, by definition of δ as the dual operator with respect to d,

$$\int \langle \delta\omega, \alpha \rangle = \int \langle \omega, d\alpha \rangle$$

$$= \quad , \text{ using (4.4)},$$

$$\int \langle \omega, \ d\alpha' + d\beta d\tau \rangle$$

$$= \quad , \text{ using (4.2)},$$

$$\int \langle \omega_B + \omega_E d\tau, \ d\alpha' + d\beta d\tau \rangle$$

$$= \int <\omega_B + \omega_E d\tau, \; d'(\alpha') + \partial_\tau(\alpha')d\tau + d\beta d\tau>$$

$$= \int [<\omega_B, d'\alpha'> + <\omega_E, \partial_\tau(\alpha') + d\beta> <(d\tau, d\tau>]$$

= , using (4.1),

$$\int [<\omega_B, d'\alpha'> - <\omega_E, \partial_\tau(\alpha') + d\beta>]$$

$$= \int <\delta'\omega_B, \alpha'> + \int <\partial_\tau(\omega_E), \alpha'> - \int <\delta'\omega_E, \beta>$$
(4.6)

where "δ" is the dual on the force perpendicular to $d\tau$. We are also assuming that

$$\partial_\tau$$

is skew-adjoint, which is an additional condition on the time-like function τ. (In fact, it requires that

grad τ

be a *Killing vector field*.)

We these assumptions we have proved:

<u>Theorem 4.1</u>. The Maxwell equation

$$\delta\omega = J$$

$$\equiv J_B + J_E \, d\tau$$

is equivalent to the following set:

$$d'\omega_B = 0$$

$$\partial_\tau(\omega_B) + d\omega_E = 0$$

$$\delta'\omega_B + \partial_\tau(\omega_E) = J_B$$

$$\delta'\omega_E = -J_E ,$$
(4.7)

which is the usual three-dimensional version of Maxwell's equations.

Maxwell's Equations 181

5. KINETIC AND POTENTIAL ENERGY IN NEWTONIAN MECHANICS

In preparation for the study of "energy" in electromagnetic theory, let us review its place in Newtonian particle mechanics. Consider a mechanical system with one degree of freedom, i.e., a one-dimensional configuration space. Let q denote the *position* coordinate, v the *velocity* coordinate. (q,v) together constitute the *tangent bundle* to configuration space, or the *state space* in the sense of statistical mechanics and system theory. Newtonian Lagrangians are functions of these variables of the following form:

$$L(q,v) = \frac{1}{2} mv^2 - U(q)$$

$$= T - U .$$

T is, of course, the *kinetic*, U the *potential* energy. $E = T + U$ is the *total* energy.

These ideas far transcend mechanical systems of one degree of freedom. q can be a point in a real vector space Q. m can be a quadratic form on Q. (q,v) can denote a point in $Q \times Q$. L is a real-valued function on $Q \times Q$.

$$L(q,v) = \frac{1}{2} m(v,v) - U(q)$$

$$= T - U .$$

6. KINETIC AND POTENTIAL ENERGY IN A RELATIVISTIC FORM

The material in the proceeding sections is *non-relativistic*, since space and time play a separate role. The basic geometric objects are curves

$$\underline{q}: t \to q(t)$$

in configuration space.

One can put space and time on a more equal footing. Associate to the curve $\underline{q}$ its *graph*

$$\underline{q}': gr(\underline{q}): t \to (t, q(t))$$

This is a curve in $T \times Q$ where T is the time interval. The action function

$$L(\underline{q}) = \int \frac{1}{2} m \left(\frac{dq}{dt}\right)^2 - U(q)$$

can be written as the integral

$$\int \left(\frac{1}{2} m\dot{q}^2 - U(q)\,t \right) \quad,$$

i.e., as the action associated to the Lagrangian

$$L' = \frac{1}{2} m\dot{q}^2 - U(q)\,t$$

in $Q \times T$. This is the homogenization process. It enables us to recognize kinetic and potential energy for Lagrangians given in relativistic form.

7. KINETIC AND POTENTIAL ENERGY FOR MAXWELL FIELDS

Let X be a Riemannian manifold. Let ω be a two differential form on X,

$$L(\omega) = \int <\omega,\omega> \tag{7.1}$$

is the *action*.

Let τ be a one-differential form. ω can be decomposed as follows:

$$\omega = \omega_B + \omega_E \tau \quad, \tag{7.2}$$

where ω_B is a two-form, ω_E a one-form, which belongs to the Grassmann algebra of forms generated by those which are perpendicular to τ. Then

$$L(\omega) = -\int <\omega_B,\omega_B> + \int <\omega_E,\omega_E> \quad .$$

Define the *kinetic* and *potential* energy as follows:

$$KE(\omega) = \int - <\omega_B,\omega_B>$$

$$PE(\omega) = \int <\omega_E,\omega_E> \quad .$$

Just as in Newtonian particle mechanics, the "action", i.e., the quantity to be extremized to give the equations of motion, is the difference of the kinetic and potential energy. However, notice a difference from Newtonian particle mechanics and the kinetic and potential energy are on a more equal mathematical footing, and indeed, can be "mixed up" by *Lorentz transformations*.

If we specialize the underlying manifold X to be R^4, space-time, and the metric to be the Lorentz metric, we obtain the familiar formulas to be found in all the physics books.

Maxwell's Equations

Let (x,y,z,t) be the usual space-time coordinates of X and let

$$ds^2 = -c^2 dt^2 + dx^2 + dy^2 + dz^2$$

be the Lorentz metric. Let

$$\tau = d\tau .$$

Suppose ω is a two-differential form on X. Let

$$\omega = \omega_B + \omega_E \, dt$$

be its *Maxwell decomposition*. Thus, ω_B and ω_E are differential forms (ω_B of degree 2, ω_E of degree 1) in dx, dy, dz with coefficients which are functions of (x,y,z,t). (Physicists would call ω_E a time-dependent *vector field*, ω_B a time-dependent *axial vector field*.

Let

$$\partial_\tau(\omega_B), \; \partial_\tau(\omega_E)$$

be the forms which result from differentiating the coefficients of these forms with respect to t. Let

$$d'\omega_B, \; d'\omega_E$$

be the forms obtained by applying three-dimensional exterior differentiation to the forms *with* t *held constant*. ω_B is the magnetic, ω_E the electric field. Thus,

$$d\omega = d'\omega_B + \partial_\tau(\omega_B)dt + d'\omega_E \, dt .$$

The first of Maxwell's equations then implies that:

$$d'\omega_B = 0$$

$$\partial_\tau(\omega_B) + d'\omega_E = 0 .$$

The action is:

$$\int <\omega_B, \omega_B> = \frac{1}{c^2}<\omega_E, \omega_E> .$$

Thus, the energy is:

$$KE = \int <\omega_B, \omega_B>$$

$$PE = \int <\omega_E, \omega_E>$$

If ω_B and ω_E are exhibited as vector fields B and E in the usual way, this takes the more customary form:

$$KE = \iiint |B|^2 \, dxdydz$$

$$PE = \iiint |E|^2 \, dxdydz \quad .$$

These formulas (which I believe are due to Maxwell) are basic to circuit theory. When they are "discretized" (engineering term: "lumped parameters") they give what amounts to a version of *Lagrange's equations* for circuits. I believe this forms the inspiration of the work of Kron and the RAAG memoirists.

8. FORCE IN NEWTONIAN PARTICLE MECHANICS IN TERMS OF THE CALCULUS OF VARIATIONS

We all learn

$$\text{Force} = \text{mass} \times \text{acceleration}$$

in school. However, it is difficult to sort out the true "geometric" nature of these terms! (Indeed, Einstein earned part of his fame by refining our understanding of this equation!) Now, one way of making this more precise is to start with the calculus of variations, and "derive" Newton's laws. While this is no doubt of little philosophical importance (as it was considered in the 18th century!) it is a convenient mathematical peg on which to anchor the physical ideas.

Again, consider a mechanical system of one-degree of freedom. Consider a Lagrangian of the form:

$$L = \frac{1}{2} mv^2 - U(q) \tag{8.1}$$

$$v = \frac{dq}{dt} \quad .$$

The first variation is:

$$\delta \underline{L}(\underline{q}) = \delta \int \frac{1}{2} m \left(\frac{dq}{dt}\right)^2 - U(q(t))$$

$$= m\frac{dq}{dt}\frac{d}{dt}\delta q - U_q \delta q$$

= , after integrating by parts and throwing away the boundary terms in δq,

$$-\int \left(m\frac{d^2q}{dt^2} + U_q\right)\delta q \quad .$$

The first variation is then:

$$m\frac{d^2}{dt^2} + U_q \quad ,$$

which is just the "force".

This suggests a way of developing a notion of "force" in a Cartanian framework. Let θ be the action differential form, and $\phi_1, \ldots, \phi_n$ the constraints. θ is to be chosen to satisfy relations of the form:

$$d\theta = \alpha_1 \wedge \phi_1 + \cdots + \alpha_n \wedge \phi_n$$

The equations of the extremals would be

$$\alpha_1 = \phi_1 = \cdots = \alpha_n = \phi_n = 0 \quad ,$$

in the absence of forces.

In order to see how to insert forces, i.e., as "first variation", suppose V is a vector field in X which generates a variation

$$\mathscr{L}_V(\theta) = i(V)(d\theta) + d(i(V)\theta) \quad .$$

The first variation is then

$$i(V)(d\theta) \quad .$$

Let us suppose that the trajectory curve of the physical system is a curve $t \to \sigma(t)$ in X. The *force* is a one-differential form F on X. Newton's equation is then:

$$\boxed{i(\sigma'(t))(d\theta) = F(\sigma(t))}$$

PREFACE TO CHAPTER 12

In the late 1950's, Kalman and Bucy changed the "paradigm" in the theory of filtering and prediction of stochastic processes -- and hence much of the mathematical side of electrical engineering -- by emphasizing the *matrix Riccati* ordinary differential equation in an algorithmic way to derive the *optimal recursive filter*. This fundamental advance was based on two general mathematical (or "system theoretic") features:

1) The description of a linear stochastic system (i.e., in classical engineering language a *signal* plus *noise*) in *state space* form.

2) The reduction of the optimal filter to solving a linear stochastic differential equation

The third feature that, combined with the first two, led to the great practical success of their work was the following one:

3) A fundamental solution (in physicist's language, a "Green's function") for the diffusion (or Fokker-Planck) operator associated with a linear stochastic differential equation can be constructed from a solution of the general solution of an ordinary differential equation of *matrix Riccati* type.

Another geometric principle linked filtering of linear stochastic differential equations to *deterministic* Optimal Control Theory.

4) The matrix Riccati equation is the Hamilton-Jacobi equation associated to a linear optimal control problem, i.e., Lagrange problem of the calculus of variations, in classical language. Their solutions are also determined by certain Lagrangian submanifolds of certain symplectic structures (in classical language, the "Problem of Pfaff".)

Here is another area:

5) The Matrix Riccati Equations are what Lie and Vessiot called *Lie systems*, i.e., systems of ordinary differential equations whose general solutions can be obtained after prolongation by solving a set of left-invariant differential equations on a Lie group.

We see that there is a potentially very rich area of contact between differential geometry, stochastic systems, differential geometry and Lie theory, optimal control theory, symplectic differential geometry ("symplectic engineering", as Sternberg calls it), and so on.

The geometric study of matrix Riccati equations is one of the topics on which Clyde Martin and I have collaborated. The study of the time-invariant matrix Riccati equation is geometrically equivalent to the study of the orbits of the action of one parameter subgroups of the Lie group $Sp(n,R)$ on the homogeneous space

$$Sp(n,R)/V(n) \quad ,$$

the so-called *Lagrange Grassman* manifold. (It is space of n-dimensional linear subspaces of R^{2n}, which annihilate the canonical symplectic form, the so-called *Lagrangian subspaces*.) From this point of view, what is important are the study of the *fixed points* and *periodic orbits*.

I wrote the following paper as a by-product of our collaborative work. (Lie and Morse theory for periodic orbits of vector fields and matrix Riccati equations, I and II, *Math. Systems Theory* 15, 277, 284 (1982); 16, 297-306 (1983).) There is also considerable related material in *Interdisciplinary Mathematics*, Vol. 20.

Chapter 12

PERIODIC SOLUTIONS FOR THE MATRIX RICCATI EQUATION VIA LIE THEORY

1. INTRODUCTION

The matrix Riccati equation is one of the fundamental objects of present day control theory. It is an ordinary differential equation of the following form:

$$\frac{dP}{dt} = f(P) , \qquad (1.1)$$

where P denotes an $n \times n$ real matrix, and the function f on the right hand side is a polynomial in the entries of P of degree two or less.

These ordinary differential equations can be described in terms of Lie theory. There is a subgroup H of the Lie group $G = Sp(n, R)$ (the group of real, $2n \times 2n$ matrices which, acting on R^{2n}, preserves the natural symplectic structure) such that the coset space $M = G/H$ has the following properties:

a) G/H may be identified with the so-called "Lagrange-Grassmann manifold", i.e., the manifold of n-dimensional linear subspaces of R^{2n} on which the symplectic form vanishes.

b) G/H has a dense open subset which is diffeomorphic to the space of all real $n \times n$ symmetric matrices, in such a way that the natural action of G on G/H goes over to linear fractional transformations of the space of symmetric matrices.

c) The vector field defined by the right hand side of Equation (1.1) goes over into a vector field on M which arises from the Lie algebra of G. Thus, the solutions of the ordinary differential equations (1.1) may be written down by means of explicit formulas once the one-parameter subgroups of G are described explicitly.

In control theory one is interested in obtaining various sorts of qualitative information about the nature of the solutions of (1.1). The most important piece of information is the nature of the time-invariant solutions, i.e., the symmetric matrices P such that:

$$f(P) = 0 . \qquad (1.2)$$

Now, this can be handled independently of Lie theory [1]. However, the next question of this sort, the nature of the periodic solutions of the matrix

Riccati equations, does not seem to have been described in the current literature. Our goal here is to develop geometric techniques for the description of at least some of these periodic solutions.

There is one case where one can say quite a bit qualitatively about the nature of the solutions of the matrix Riccati equations. In case the $2n \times 2n$ matrix corresponding to the element of the Lie algebra of G which generates the solutions of (1.1) has real eigenvalues, one can prove [2] that there is a Riemannian metric on G/H and a real-valued function such that the solution curves of (1.1) are gradient curves of this function relative to the Riemannian metric. This enables one to use Morse theory [3] to derive considerable information about the qualitative nature of the time-independent solutions of (1.1). (Further, in this case, there are no periodic solutions.) Using results of Palis this also implies useful information about the case where the eigenvalues are sufficiently close to being real, and are distinct. The group acts on the compact manifold G/H in a so-called Morse-Smale way. This property has certain implications about the qualitative nature of the solutions of (1.1).

The aim of this paper is to discuss certain periodic solutions of the matrix Riccati equation using Lie theoretic techniques.

2. CERTAIN TYPES OF PERIODIC ORBITS AS FIXED POINTS.

Let X be a manifold and let A, B, and C be vector fields of X such that:

$$C = A + B \qquad (2.1)$$

$$[A,B] = 0 \qquad (2.2)$$

and

> A, B, and C are complete, in the sense that they generate one-parameter groups of diffeomorphisms of the manifold X.

Let $t \to \exp(tA)$ denote the group of diffeomorphisms generated by the vector field A. As a consequence of (2.1) and (2.2) we have:

$$\text{ext}(tC) = \exp(tA)\exp(tB) \qquad (2.3)$$

for all real t.

Recall that an orbit $t \to \exp(tC)(x)$ of the group generated by C is said to be *periodic* if there is a real number T such that:

Periodic Solutions

$$\exp((t+T)C)(x) = \exp(tC) \qquad (2.4)$$

for all real t.

In view of the group property, the condition for (2.4) to hold is that:

$$\exp(TC)(x) = x, \qquad (2.5)$$

i.e., that the point x is a fixed point for the diffeomorphism $g = \exp(TC)$. Of course, (2.5) is implied by the condition that x be a zero point of the vector field C, i.e.,

$$C(x) = 0.$$

An orbit which arises from such a zero point will be called a trivial periodic orbit.

Assuming (2.2), our goal is to derive conditions for existence of non-trivial periodic orbits of C in terms of conditions on the orbits of A and B. In fact, we shall work here only with the most obvious of such conditions, namely, the one provided by the following result.

Theorem 2.1. With the notation described above, suppose that the point $x \in X$ has the following properties:

a) The orbit $t \to \exp(tB)(x)$ is periodic, $\qquad (2.6)$

b) $A(x) = 0$. $\qquad (2.7)$

Suppose also that conditions (2.1) and (2.2) are satisfied. Then, the orbit of C passing through x is periodic.

The proof is obvious from condition (2.2).

Let us continue to suppose that conditions (2.1) and (2.2) are satisfied. Pick a real number T. Set:

$$g = \exp(TB).$$

Then, g is a diffeomorphism of X. Set:

$$Y = \text{fixed point set of } g$$
$$= \{x \in X: gx = x\}.$$

Theorem 2.2. Suppose that the fixed point set Y is a regularly embedded submanifold of X. Then A and B are both tangent to Y. Let Y_0 be the subset of points of Y at which the vector field B is zero. Then, the points of $Y - Y_0$ which are zero points of A are points which lie on non-trivial periodic orbits of the vector field $C = A + B$.

Again, the proof should be obvious.

We now plan to use these results to generate a "strategy" for finding nontrivial periodic orbits of certain types of vector fields. The situation is simplest if A and B are part of a Lie algebra of vector fields on X which arise from a transitive transformation group action of the Lie group G. This is the situation we now plan to study.

3. A LIE-THEORETIC SITUATION

Let G be a connected Lie group and let H be a closed Lie subgroup. Let $X = G/H$ be the coset space, with the natural transformation group action of G on X. Pick a point $g \in G$. Set

$$Y = \{x \in X: gx = x\} \ .$$

Theorem 3.1. Suppose that the following condition is satisfied:

$$\text{Ad } g, \text{ acting in the Lie algebra } \mathcal{G}, \text{ is completely reducible.} \tag{3.1}$$

Then, Y is a closed, regularly embedded submanifold of X.

The proof of this result may be considered as Lie-theoretic "folklore".

Let us now set:

L = centralizer of g in G .

$\mathcal{L}$ = Lie algebra of L .

Consider $\mathcal{G}$, the Lie algebra of G, as a Lie algebra of vector fields on X. $\mathcal{L}$ is then the tangent of the submanifold Y.

Theorem 3.2. L acts transitively on each connected component of Y.

Theorem 3.3. Suppose that B is an element of $\mathcal{L}$ such that:

$$g = \exp(B) \ . \tag{3.2}$$

Suppose that A is an element of L such that:

$$[A,B] = 0 \ .$$

Then, the points y of Y which are zero points of A, but are not zero points of B, are nontrivial periodic orbits of $C = A + B$.

4. THE CASE WHERE G/H IS A COSET SPACE SUCH THAT (G,H) IS A GRASSMANN PAIR

Keep the notation of Section 3. In addition, suppose that the following conditions are satisfied:

- G is semi-simple
- X = G/H is compact
- K, a maximal compact subgroup of G, act transitively on X
- g ∈ K .

With these conditions satisfied, we can put a Riemannian metric on X which is invariant under the action of K. The following theorem proved by Kobayashi is very relevant.

<u>Theorem 4.1.</u> Y is a totally geodesic submanifold of the Riemannian manifold X.

The following result is proved in [3]:

<u>Theorem 4.2.</u> If a vector field V on Y is a gradient vector field relative to a Riemannian metric on X, if V is tangent to the sum Y, then V restricted to Y is a gradient vector field relative to the Riemannian metric on Y induced by the Riemannian metric on X.

<u>Definition.</u> Let G be a Lie group and let H be a subgroup. We shall say that the pair (G,H) is a *Grassmann pair* if there is some linear action of G on a complex Grassmann manifold (defined via a linear representation of G) such that H is the isotropy subgroup at one point of the Grassmann manifold. In this case, H is closed, and the coset space G/H is identified with a submanifold of a Grassmann manifold.

<u>Theorem 4.3.</u> Let G be a connected Lie group and let H be a Lie subgroup of G such that H is the normalizer of $\mathcal{H}$ in $\mathcal{G}$, i.e., the following is satisfied:

$$H = \{g \in G : \operatorname{Ad}(g)(\mathcal{H}) = \mathcal{H}\} . \qquad (4.1)$$

Then, the pair (G,H) is a Grassmann pair, in the sense of the above definition.

<u>Proof.</u> Use the adjoint linear representation of G to let G act on the Grassmann manifold of m-dimensional linear subspaces of the vector space $\mathcal{G} \times C$, where m = dim H. It follows from (4.1) that G/H is the orbit of this action of G which passes through the point of the Grassmann manifold determined by $\mathcal{H} \times C$.

Here is a useful condition, which is proved in [2], that a vector field be a gradient vector field.

<u>Theorem 4.4.</u> Let G be a connected, semi-simple Lie group and let H be a closed Lie subgroup such that (G,H) is a Grassmann pair, in the sense of the above definition. Suppose also that A is an element of the Lie algebra $\mathcal{G}$ which is perpendicular (with respect to the Killing form) to some maximal compact subalgebra $\mathcal{K}$ of $\mathcal{G}$. Then, there is a Riemannian metric on G/H that is invariant under the action of K such that A, considered as a vector field of G/H, is a gradient vector field with respect to the Riemannian metric and some function on G/H.

Now, let us choose G as a connected, semi-simple, non-compact Lie group whose center is finite. Let K be a maximal compact subgroup of G. Let H be another closed subgroup of G. Let X be the coset space G/H. Suppose that the following condition is satisfied:

$$\text{K acts transitively on X .} \qquad (4.2)$$

X is then also exhibited as the coset space K/L, where L is the intersection of the groups K and H.

Let g be an element of L and let $\mathcal{P}$ be the orthogonal complement of $\mathcal{K}$ in $\mathcal{G}$ with respect to the Killing form of $\mathcal{G}$. Then:

$$\text{Ad}(g)(\mathcal{P}) = \mathcal{P} .$$

Let x_0 be the identity coset element of X = G/H = K/L. Let K' be the subgroup of K consisting of the elements $k \in K$ such that:

$$gk = kg ,$$

i.e., K' is the centralizer of g in K.

<u>Theorem 4.5.</u> The orbit of K' at x_0 is a connected component of the set of fixed points of g acting on X. This orbit is the coset space

$$\text{K'/(centralizer of g in L) .}$$

Again, the proof follows readily from general Lie theoretic principles.

We can now describe some of the periodic orbits of one-parameter subgroups of G, acting on X = G/H. Let g be a fixed element of L, as described above. Choose an element B of $\mathcal{L}$ such that:

$$\exp(B) = g .$$

Let $\mathcal{P}$ be the orthogonal complement of $\mathcal{K}$ in $\mathcal{G}$ with respect to the Killing form of $\mathcal{G}$. Choose an element $A \in \mathcal{P}$ such that:

$$[A,B] = 0 .$$

A, considered as a vector field on X, is the gradient of a function with respect to the Riemannian metric on X, which is invariant under the action of K. Let X' be the fixed point set of g acting on X. Since A commutes with g, A is tangent to the submanifold X'. Let A' denote this vector field restricted to X'. Let f' denote the function f restricted to X'. A' is then the gradient of f' with respect to the K-invariant Riemannian metric restricted to X'. We can then put all of this together to prove the following main result:

<u>Theorem 4.6.</u> With the above notation, the critical points of f' on the compact manifold X' define periodic orbits of the vector field C = A + B. Those critical points which are not zero points of B define nontrivial periodic orbits.

5. PERIODIC ORBITS OF CERTAIN VECTOR FIELDS ON COMPACT HOMOGENEOUS SPACES

We can now generalize the constructions given in the previous section. Let K be a connected compact Lie group, and let L be a closed subgroup. Let X be the coset space K/L. The Lie algebra $\mathcal{K}$ then acts as a Lie algebra of vector fields on X. Fix a Riemannian metric on X which is invariant under the action of K. For f ∈ F(X), let grad(f) denote the gradient of f with respect to the Riemannian metric. Let g be an element of K and let X' be the fixed point set of g acting on X. Let B be an element of $\mathcal{L}$ such that:

$$g = \exp(B) \, .$$

Let f be an element of F(X) such that:

$$B(f) = 0 \, . \tag{5.1}$$

<u>Theorem 5.1.</u> With the above conditions, the critical points of f restricted to the submanifold X' are points that lie on periodic orbits of the vector field B + grad(f).

<u>Proof.</u> Condition (5.1), plus the fact that B is a Killing vector field with respect to the Riemannian metric, i.e., generates a one-parameter group of isometries, guarantees that:

$$[B, \text{grad}(f)] = 0 \, .$$

The rest of the proof is as before.

This result suggests that we look for ways of satisfying condition (5.1). We are interested not only in choosing such an isolated f, but one belonging

to a whole family on which K acts. This suggests that we choose the f
from a linear family of functions on X that transform under K via an
irreducible linear representation. To see how this may be done, we can invoke
the Frobenius Reciprocity Theorem in its "vector bundle" version. (The treatment in [LPG] is the most convenient from our point of view.) The "natural"
action (without multipliers) of K in F(X) is the representation induced
from L to K by choosing the identity representation of L. Thus, a given
irreducible representation ρ of K appears as many times in the action of
K on F(X) as ρ restricted to L contains the identity representations.
Of course, this property ties in directly with our interest in constructing
vector fields which will have periodic orbits by the method described above.
Here is the main example of interest in the theory of the matrix Riccati
equation.

6. CLASSES OF VECTOR FIELDS ON THE LAGRANGE-GRASSMANN MANIFOLD THAT HAVE PERIODIC ORBITS

Continue with the notation of Section 5. However, let us now specialize
as follows:

$$K = U(n)$$

$$L = O(n,R) .$$

$$X = K/L = \text{the Lagrange-Grassmann manifold.}$$

Let $G = Sp(n,R)$. G also acts on X, hence its Lie algebra $\mathcal{G}$ acts as a
Lie algebra of vector fields on X. Let $\mathcal{P}$ be the orthogonal complement of
$\mathcal{K}$ in $\mathcal{G}$ with respect to the Killing form of $\mathcal{G}$. Let ρ be the representation of Ad K in $\mathcal{P}$. (Since G/K is an irreducible symmetric space [$\mathcal{HELG}$
ρ is an irreducible representation.) $\mathcal{P}$ may be identified with the $n \times n$
Hermitian matrices, with ρ just the natural action

$$h \to UhU^{-1}$$

of unitary matrices on the Hermitian ones. We see that $\rho(K)$ does leave
invariant a one-dimensional linear subspace of $\mathcal{P}$, namely, that generated
by the identity matrix. Thus, we have a confirmation of the fact that $\mathcal{P}$
acts on X via gradient fields; namely, the following result holds.

Theorem 6.1. In K acting on F(X), there is exactly one linear subspace of
F(X) in which K acts via the irreducible representation ρ. The functions
in this subspace, when converted into gradient vector fields using the
Riemannian metric on X, which is invariant under K, generate the Lie
algebra $\mathcal{P}$ of vector fields acting on X.

Periodic Solutions

Theorem 6.2. Let g be an element of the real orthogonal group O(n,R). Suppose that n - 2m of the eigenvalues of g are equal to one, while the rest are distinct. Let g act on the coset space U(n)/O(n,R). Then, the connected component of the fixed set of g is the coset space:

$$T^m \times \frac{U(n-2m)}{O(n-2m, R)} ,$$

i.e., a product of circles (and a circle is itself a Lagrange-Grassmann manifold) and a Lagrange-Grassmann manifold.

We can now extent to the case where g does not have distinct eigenvalues.

Theorem 6.3. With the hypotheses of Theorem 6.2., suppose that g is an orthogonal manifold whose eignevalues are not necessarily distinct. Then, the connected component of the fixed set of g acting on U(n)/O(n,R) is a product of Lagrange-Grassmann manifolds.

Proof. Just a refinement of the argument. Label the eignevalues so that equal ones are adjacent. Diagonalize the matrix g. Anything commuting with g must ghen preserve the eigenvectors for the same eigenvalue. The structure of the fixed set follows.

We can now see how periodic orbits of matrix Riccati equations may be constructed. Let X continue as the coset space K/L, with K = U(n), L = O(n,R). Let $\mathcal{P}$ be the vector space of n × n Hermitian manifolds. Let ρ be the natural action of K in $\mathcal{P}$. ρ is an irreducible representation of K. ρ restricted to L has precisely one independent fixed vector, namely, the identity matrix. By Frobenius reciprocity, there is a unique (up to a scalar multiple) linear map α: $\mathcal{P} \to F(X)$ which intertwines the action of K. Since K/L is an irreducible Riemannian symmetric space, there is a unique (up to scalar multiple) Riemannian metric on X that is invariant under the action of K. The gradient operation (with respect to this invariant Riemannian metric) is a K-intertwining linear map: $F(X) \to \mathcal{V}(X)$; thus, grad(α($\mathcal{P}$)) is a linear subspace of $\mathcal{V}(X)$ which is invariant under the action of K, and in which K acts via the representation ρ.

Now, we are interested in finding pairs (B,A) ∈ $\mathcal{K} \times \mathcal{P}$ with the following properties:

$$[A,B] = 0 ,$$

$$g = \exp(B), \text{ acting on } X, \text{ has a fixed point.}$$

Since B is a skew-Hermitian matrix, we can diagonalize it. It has eigenvalues that we may label as $i\lambda_1, \ldots, i\lambda_n$, where the λ's are real numbers. g has eigenvalues $\exp(i\lambda_1), \ldots, \exp(i\lambda_n)$. In order that g, acting on $C^n = R^{2n}$ admit an n-dimensional invariant subspace, i.e., in order that g acting on the Lagrange-Grassmann manifold have a fixed point, these eigenvalues must appear in complex conjugate pairs. Thus, we can normalize the labelling of the eigenvalues so that they are of the following form:

$$\exp(\pm i\lambda s j), \cdot 1,$$

where $j = 1,\ldots,m$ and $n - 2m$ of the eigenvalues are equal to one.

Let K' be the centralizer of g in K. Any g' ∈ K' then must be a direct sum of a $2m \times 2m$ diagonal matrix (unitary, of course), and an arbitrary $n - 2m \times n - 2m$ unitary matrix. Thus, K' has the following structure:

$$T^{2m} \times U(n - 2m) \quad,$$

where T^{2m} denotes the 2m-dimensional torus. The isotropy subgroup of K' acting on X is then the group

$$T^m \times O(n - 2m) \quad.$$

The coset space of the first group listed above by the second is now the connected component of the identity coset of the fixed points of g acting on X. We can then sum up as follows:

We can now choose B as an element of $\mathcal{K}$. Without loss in generality, we can suppose that B is a diagonal matrix, i.e., that B belongs to one fixed Cartan subalgebra of the Lie algebra $\mathcal{K}$. We also suppose that:

$$g = \exp(B) \in L = O(n,R) \quad.$$

This condition requires that the eigenvalues of B occur in complex conjugate pairs modulo 2π, i.e., that they be of the following form:

$$2\pi i k_1, \ldots, 2\pi i k_m, \, 2\pi i k_{m+1} \pm i\xi_1, \ldots \quad,$$

with integer k's and nonzero ξ's between 0 and 2π.

Let us now choose $f \in \alpha(\mathcal{P})$ so that:

$$B(f) = 0 \quad.$$

Set:

$$C = B + \text{grad } f \quad.$$

We see that

$$[B, \text{grad } f] = 0 \quad.$$

Let K' be the centralizer of $g = \exp(B) \in L$, and let L' be the centralizer of g in L.

Theorem 6.4. The critical points of f restricted to K'/L' are non-trivial periodic orbits of C.

References

1. J.C. Willems, Least squares stationary optimal control and the algebraic Riccati equation, *IEEE Trans. Auto. Control* AC-$\underline{16}$, 621-634 (1971).

2. R. Hermann, Geometric aspects of potential theory in symmetric spaces, I-III, *Math Ann.* $\underline{148}$, 349-366 (1962); $\underline{151}$, 384-395 (1964).

3. J. Milnor, *Morse Theory*, Princeton Univ. Press, 1963.

4. R. Hermann, Compactification of homogeneous spaces, I-II, *J. Math. Mech.* $\underline{14}$, 655-678 (1965); $\underline{15}$, 667-681 (1966).

5. R. Hermann, Compactification of homogeneous spaces and contractions of Lie groups, *Proc. Nat'l Acad. Sci.* $\underline{51}$, 456-461 (1964).

6. S. Kobayashi, Fixed points of isometrics, *Nagoya Math. J.* $\underline{13}$, 63-68 (1958).

7. S. Helgason, *Differential Geometry and Symmetric Spaces*, Academic Press, 1961.

8. A. Borel, Sur la cohomologie des espaces filtres principaux et des espaces homogenes des groupes de Lie compacts, *Annals of Math.* $\underline{57}$, 115-176 (1953).

PART IV

STOCHASTIC SYSTEMS

The next two chapters are a small beginning toward a theory of integrable stochastic systems. Chapter 13 was written four years ago, about the time of the NATO Les Arcs Conference (which I did not attend) when the control theory community was excited about the possibility of generalizing the Kalman-Bucy filter via Lie algebra theory. I believe that the ideas proposed then were too simplistic, and basically old-hat in an analogous quantum mechanical context. Tyrone Duncan, as usual, could see furthest: The road to useful generalization of the Kalman-Bucy filter leads via a deeper study of the underlying geometric structures, and its associated infinite dimensional (e.g., Kac-Moody) Lie algebras. I am writing more material myself about this, which might appear in Volume 24 or 25.

Chapter 13

DIFFERENTIAL GEOMETRY AND LIE THEORY OF CLASSICAL AND QUANTUM STOCHASTIC SYSTEMS.

1. INTRODUCTION

Stochastic system theory concerns itself with Ito differential equations of the form

$$dx = f(x)dt + g(x)dw_1$$
$$dy = h(x)dt + dw_2 \quad . \tag{1.1}$$

It is well-known that these systems are equivalent to partial differential equations of evolution type:

$$\frac{\partial \rho}{\partial \tau} = \frac{\partial}{\partial x}(f(x)\rho) + \frac{\partial^2 \phi}{\partial x^2} + \frac{\partial}{\partial y}(h(x)\rho) + \frac{\partial^2 \rho}{\partial y^2} \quad . \tag{1.2}$$

(These are the equations determining the probability densities associated with the random variable solutions of (1.1).)

The aim of this chapter is to reformulate Equations (1.2) in a coordinate free, manifold setting, and to study some Lie theoretic aspects of these systems.

A new feature of our treatment is the introduction of the point of view of the classical Galois-Picard-Vessiot theory of factorization of differential operators. We hope to use it to generalize the *annihilation-creation operator formalism* of quantum mechanics, and to give a more precise *algebraic* setting to recent attempts to generalize the Kalman-Bucy filter.

This is a theory with many scientific and mathematical ramifications, hence I will present certain background material that will be necessary to realize its potential for cross-disciplinary insight.

We shall formulate the concepts as far as possible in terms of coordinate-free manifold theory. (Of course, I will not be fanatical about this. Many examples are best presented in the traditional notation used by engineers and physicists.) In order not to hinder the flow of these geometric ideas and the development of the algebraic formalism, we shall work in the context of smooth (i.e., C^∞) geometric objects. Methods are available (e.g., de Rham's theory of currents) for extending to non-smooth situations. Similarly, development of quantum mechanical ideas requires functional analysis techniques. Here, we shall follow the physicists, emphasizing the algebraic aspects and working "formally", i.e., ignoring such issues as domains of operators, topologies, continuity of mappings, etc.

Let X be a manifold. $\mathscr{D}^n(X)$ denotes the n-th degree differential form. $\mathscr{D}^n(X) = \mathscr{F}(X)$ are the (C^∞) real-valued functions on X.

$$d: \mathscr{D}^n(X) \to \mathscr{D}^{n+1}(X)$$

denotes exterior derivative. $\mathscr{V}(X)$ denotes the vector fields on X. For $V \in \mathscr{V}(X)$, $\theta \in \mathscr{D}^n(X)$,

$$V \lrcorner \theta$$

denotes the contraction of θ by V, while

$$\mathscr{L}_V(\theta)$$

denotes *Lie derivative*.

2. DIFFUSION EQUATIONS ON MANIFOLDS

Let X be a manifold of dimension n. A linear operator

$$D: \mathscr{D}^n(X) \to \mathscr{D}^n(X)$$

will be called a *diffusion operator* if it is of the following form:

$$D = d\delta \quad , \tag{2.1}$$

where

$$\delta: \mathscr{D}^n(X) \to \mathscr{D}^{n-1}(X) \tag{2.2}$$

is a linear differential operator. (See [19] for a coordinate-free definition of "linear differential operator" in the context of cross-sections of vector bundles.) Given such an operator, the associated *diffusion equation* is

$$\frac{\partial \phi}{\partial \tau} = D\rho \quad , \tag{2.3}$$

to be solved for a curve $t \to \rho(t)$ in $\mathscr{D}^n(X)$. Modulo some conditions on the behavior at infinity, Stokes' formula then guarantees that

$$\frac{d}{dt}\left(\int_X \phi(t)\right) = 0 \quad , \tag{2.4}$$

i.e., "probability" or "mass" is conserved.

The operator is said to be *deterministic* if δ is of the following form:

$$\delta(\rho) = V \lrcorner \rho \quad , \tag{2.5}$$

for a vector field $V \in \mathscr{V}(X)$. In this case, the operator D is also of the following form:

$$D(\phi) = \mathscr{L}_V(\rho) \quad . \tag{2.6}$$

Equation (2.2) is then called *Liouville's equation* in the physics literature.

Definition. The diffusion operator

$$d\delta : \mathscr{D}^n(X) \to \mathscr{D}^n(X)$$

is said to be of *Fokker-Planck type* (abbreviation: FP) if it is of the following form:

$$\delta(\rho) = V \lrcorner \rho + \omega(\rho) \quad , \tag{2.7}$$

with $V \in \mathscr{V}(X)$ and

$$\omega : \mathscr{D}^n(X) \to \mathscr{D}^{n-1}(X) \tag{2.8}$$

a *first order* linear differential operator.

Let us compute the commutator of two operators $d\delta$, $d\delta'$ of this type:

$$[d\delta, d\delta'] = d\delta d\delta' - d\delta' d\delta$$

$$= d(\delta d\delta' - \delta' d\delta) \tag{2.9}$$

Let us now suppose that $d\delta$ is deterministic, i.e.,

$$d\delta = \mathscr{L}_V \quad , \tag{2.10}$$

with $V \in \mathscr{V}(X)$. Then,

$$[d\delta, d\delta'] = [\mathscr{L}_V, d\delta']$$

$$= d\mathscr{L}_V(\delta') \quad , \tag{2.11}$$

since the exterior derivative operator d commutes with the Lie derivative operation.

In particular, let us suppose that $d\delta'$ is of FP-type, i.e.,:

$$d\delta' = \mathscr{L}_{V'} + d\sigma \quad , \tag{2.12}$$

where $\sigma: \mathcal{D}^n(X) \to \mathcal{D}^{n-1}(X)$ is of first order. Then,

$$[d\delta, d\delta'] = \mathcal{L}_{[V,V']} + d\mathcal{L}_V(\sigma) \qquad (2.13)$$

One particular goal is to study Lie algebras of differential operators, which we generate by FP and deterministic operators.

Let us now turn to the study of the analogous quantum concept.

3. QUANTUM DIFFUSION OPERATORS

Let H be a Hilbert space, i.e., a complex vector space with a positive definite Hermitian inner product

$$(\psi_1, \psi_2) \to \langle \psi_1, \psi_2 \rangle$$

Just as in the pervious section we put to the side measure theoretic subtleties, so here we ignore all functional-analytic subtleties such as domains of operators, continuity of maps, etc. L(H) denotes the space of linear maps: $H \to H$.

$\mathcal{O}$ denotes the vector space of quantum mechanical observables, i.e., the set $\rho \in L(H)$ which are Hermitian, i.e., satisfy

$$\langle \psi_1, \rho \psi_2 \rangle = \langle \rho \psi_1, \psi_2 \rangle \qquad (3.1)$$

for $\psi_1, \psi_2 \in H$

A linear operator

$$D: \mathcal{O} \to \mathcal{O}$$

is said to be a *quantum diffusion operator* if it is of the following form:

$$D(\rho) = \sum_{i=1}^{m} A_i \rho B_i \quad , \qquad (3.2)$$

for some set $(A_1, \ldots, A_m, B_1, \ldots, B_m)$ of elements of L(H), and satisfies the following conditions

$$\text{trace}(D(\rho)) = 0 \qquad (3.3)$$

for $\rho \in \mathcal{O}$

Given such an operator D, we can set up the evolution equation ("the quantum FP equation")"

$$\frac{\partial \rho}{\partial \tau} = D\rho \qquad (3.4)$$

to be solved for a curve $t \to \rho(t)$ in $\mathcal{O}$. Condition (3.3) guarantees (formally) that:

$$\frac{d}{dt} \text{trace}(\rho(t)) = 0 \quad . \qquad (3.5)$$

The significance of (3.5) for statistical mechanics is that the trace function $\rho \to \text{trace } \rho$ is conserved along solutions of (3.5). The operators ρ of trace one are then preserved--they are, of course, the quantum mechanical analogue of probability measure in classical mechanics.

Just as for the "classical" operators defined in the previous section, we can define the "deterministic" operators. They are of the form:

$$D(\rho) = [h,\rho]$$
$$= h\rho - \rho h \qquad (3.6)$$

for an $h \in L(H)$.

The corresponding evolution equation

$$\frac{\partial \rho}{\partial \tau} = [h,\rho] \qquad (3.7)$$

is called the *Heisenberg equation of motion* in the physics literature.

4. THE ALGEBRA OF THE QUANTUM MECHANICAL FP OPERATORS

Let us study the algebra generated by the operators considered in Section 2.

Quantum Stochastic Systems

Let $\mathscr{A}_1$ and $\mathscr{A}_2$ be associative algebras over a scalar field K. (K = real or complex numbers will suffice.) The tensor product

$$\mathscr{A}_1 \otimes \mathscr{A}_2$$

inherits an associative algebraic structure:

$$(A_1 \otimes A_2)(A_1' \otimes A_2') = (A_1 A_2) \otimes (A_2 A_2') \tag{4.1}$$

If $\mathscr{A}$ is an associative algebra, we can define a new associative algebra $\mathscr{A}'$ as follows:

> $\mathscr{A}'$ has the same element as $\mathscr{A}$. $A \to A'$ is the correspondence. The product in $\mathscr{A}'$ is the product in $\mathscr{A}$ *in reverse order*

$$A_1' A_2' = (A_2 A_1)' \quad . \tag{4.2}$$

<u>Proof that</u> $\mathscr{A}'$ <u>is associative</u>.

$$(A_1' A_2') A_3' = (A_2 A_1)' A_3'$$
$$= (A_3(A_2(A_1)))'$$

$$A_1'(A_2' A_3') = A_1'(A_3 A_2)'$$
$$= ((A_3 A_2) A_1)' \quad .$$

We can now define

$$\mathscr{A}_{FP} = \mathscr{A} \otimes \mathscr{A}'$$

$\mathscr{A}_{FP}$ (with the tensor product associative algebra structure) will be called the (quantum) FP-algebra. As an associative algebra, it also has a Lie algebra structure. Applied to $\mathscr{A} = L(H)$, this will be the Lie algebra structure we will need to study quantum FP-operators. For, the mapping

$$A_1 \otimes A_2 \to D(A_1, A_2) : \rho \to A_1 \rho A_2$$

is an isomorphism between $L(H) \otimes L(H)'$ and the FP operators.

Return to the case of a general associative algebra $\mathcal{A}$. Suppose it has a unit element 1. Define the *Lie mapping*

$$A \to A \otimes 1 - 1 \otimes A' \quad . \tag{4.3}$$

The following result is readily proved (and is just an abstract form of a result which is obvious for operators).

<u>Theorem 4.1</u>. The Lie mapping (4.1) is a Lie algebra homomorphism from $\mathcal{A}$ to the Lie algebra structure on $\mathcal{A} \otimes \mathcal{A}'$.

<u>Remark</u>. In view of their roles in the Liouville-Heisenberg equations, one might call the elements of $\mathcal{A} \otimes \mathcal{A}$ the *Liouville-Heisenberg* elements.

5. SOME CLASSICAL AND QUANTUM OPERATORS OF FP-TYPE

After these abstract generalities, let us turn to the simplest examples from classical and quantum mechanics. Let x be a single real variable, physically the position of a mechanical system with one degree of freedom. The simplest classical FP equations are the following, often said to be of "Langevin type".

$$\frac{\partial \rho}{\partial \tau} = a \frac{\partial}{\partial x}(x\rho) + b \frac{\partial^2 \rho}{\partial x^2} \quad . \tag{5.1}$$

This is "classical". It corresponds to an Ito-type equation of the form

$$\frac{dx}{dt} = ax + bw \quad , \tag{5.2}$$

where $t \to w(t)$ is "white noise"

$$E(w(t)y) = 0$$

$$E(w(t)w(t')) = \delta(t - t')$$

Quantum Stochastic Systems

Of course, this can be generalized to the "nonlinear" form:

$$\frac{\partial \rho}{\partial \tau} = \frac{\partial}{\partial x}(a(x)\rho) + \frac{\partial}{\partial x}\left(b(x)\frac{\partial \rho}{\partial x}\right) \qquad (5.3)$$

This corresponds to a nonlinear Ito equation--indeed, the glory of Ito's theory is that it makes rigorous and precise the correspondence between diffusion-type equations like (5.3) and *stochastic* ordinary differential equations.

Write (5.3) as

$$\frac{\partial o}{\partial \tau} = D_{det} + D_{diff} \qquad (5.4)$$

Return to the simple linear Equation (5.1). Set:

$$\partial = \frac{\partial}{\partial x}$$

$$D_{det} = a\partial x$$

$$D_{diff} = b\partial^2 \;.$$

Then,

$$[D_{det}, D_{diff}] = ab[\partial x, \partial^2]$$

$$= ab([\partial x, \partial]\partial + \partial[\partial x, \partial])$$

$$= -2ab\partial^2 = -2aD_{diff} \qquad (5.5)$$

Thus, D_{det} and D_{diff} generate a two-dimensional solvable Lie algebra. A plausible idea is that "quantum" models should realize another representation of this Lie algebra.

Let us turn to the simplest quantum mechanical situation with one degree of freedom. Let H be the Hilbert space of complex valued, C^∞, rapidly decreasing functions

$$x \to \psi(x)$$

of a real varaible x.

Let

$$\delta = \frac{d}{dx}$$

$$h = -\delta^2 + x^2 \ .$$

To construct FP-operators, we shall use integral operators A of the form:

$$(A\psi)(x) = \int a(x,y)\psi(y)\, dy$$

$$(hA\psi)(x) = \int (-a_{xx}(x,y) + x^2 a(x,y))\psi(y)\, dy$$

$$Ah\psi(x) = \int a(x,y)(-\psi_{yy} + y^2 \psi(y))\, dy$$

(Subscripts denote partial derivatives.) Integrating by parts,

$$Ah\psi(x) = \int (-a_{yy}\psi(y) + ay^2 \psi)\, dy$$

$$[h,A](\psi)(x) = \int (-a_{xx} + x^2 a + a_{yy} - ay^2)\psi(y)\, dy \ .$$

Thus, $[h,A]$ is an integral operator with kernel

$$(x,y) \to (-a_{xx} + x^2 a + a_{yy} - ay^2)(x,y) \ .$$

Thus, we can construct interesting FP operators for the harmonic oscillator of the form:

$$\frac{\partial \rho}{\partial \tau} = [h,\rho] + \sum_i A_i \rho B_i$$

$$\text{trace}(\rho) = 1 \ .$$

Let us look for ρ as an integral operator with kernal $k(x,y)$:

$$(\rho\psi)(x) = \int k(x,y,t)\psi(y)\, dy$$

$$[h,\rho](\psi)(x) = \int (-k_{xx} + x^2 k + k_{yy} - ky^2)(y)$$

$A \rho B$ is an integral operator of kernel:

$$K(x,y) = \iint a(x,z)k(z,u)b(u,y)\,dz\,du$$

Thus, the FP-equation is:

$$\frac{\partial k}{\partial \tau} = -k_{xx} + x^2 k + k_{yy} - ky^2$$

$$+ \iint \sum_{i=1}^{n} a_i(x,z)k(z,u,t)b_i(u,u)\,dz\,du$$

6. INFELD-HILL FACTORIZATION AND SOLUTION OF LINEAR EVOLUTION EQUATIONS

We have seen that FP equations--in either a "classical" or "quantum" version--leads to study evolution operators of the form:

$$\frac{d\gamma}{dt} = D\gamma, \qquad (6.1)$$

where $t \to \gamma(t)$ is a curve in a vector space Γ, and D is a linear map: $\Gamma \to \Gamma$. Now, in case Γ is a space of functions of one real or complex variable, and D is a linear differential operator with coefficients in a field of meromorphic functions, the classical Picard-Vessiot theory [20-23] describes the possible factorization of D into lower order differential operators. Certain special factorizations which are important in mathematical physics have been described in a classic paper by Infeld and Hill [24]. Some factorizations of differential operators which occur in Lie group representation theory have been described in the book by Vilenkin [25].

Let us first review what the factorization of D says about the solutions of (6.1).

Suppose that D can be factored within $L(\Gamma)$.

$$D = D''D' . \qquad (6.2)$$

Introduce the Laplace transform:

$$\underline{\gamma}(s) = \int_0^\infty e^{-st} \gamma(y) \, dt \quad . \tag{6.3}$$

Equation (6.1) becomes (formally, i.e., assuming limits can be interchanged and that integrals converge in appropriate regions),

$$s\underline{\gamma}(s) = D\underline{\gamma}(s) + \gamma(0)$$
$$= D'D''\underline{\gamma}(s) + \gamma(0) \quad . \tag{6.4}$$

Introduce another function of s

$$\underline{\gamma}'(s)$$

defined as follows:

$$\underline{\gamma}'(s) = D'\underline{\gamma} \quad . \tag{6.5}$$

Then,

$$D''\underline{\gamma}' = D''D'\underline{\gamma}$$
$$= s\underline{\gamma} - \gamma(0) \quad . \tag{6.6}$$

Equations (6.5) and (6.6) constitute a pair of equations to be solved for the curve

$$s \to (\underline{\gamma}(s), \underline{\gamma}'(s))$$

on $\Gamma \times \Gamma$. Equation (6.6) is now solved in many cases by Infeld and Hull [24].

Of course, an analogous factorization of (6.1) can be made directly in the time domain:

$$D'\gamma = \gamma' \tag{6.7}$$

$$D''\gamma' = D''D'\gamma$$
$$= \frac{\partial \gamma}{\partial t} \quad . \tag{6.8}$$

(6.7)-(6.8) constitute a pair of equations that are equivalent to the original evolution equation (6.1).

In general, we have not accomplished much in replacing one general problem with another. However, the possibility of explicit solution of Equations (6.7) and (6.8) in many problems of interest in quantum mechanics and stochastic system theory seems tied to the algebraic and geometric nature of the Lie algebra of operators generated by D' and D".

7. FACTORING SECOND ORDER DIFFERENTIAL OPERATORS IN COMMUTATIVE DIFFERENTIAL ALGEBRAS

Now, let us take an algebraic direction. Let $\mathcal{A}$ be a commutative algebra over a scalar field of characteristic zero. (The real or complex numbers will be sufficiently general.) Let

$$\delta: \mathcal{A} \to \mathcal{A}$$

be a linear map which is a derivation of the algebra structure, i.e.,

$$\delta(a_1 a_2) = \delta(a_1) a_2 + a_1 \delta(a_2)$$

$$\text{for } a_1, a_2 \in \mathcal{A} \quad .$$

The pair $(\mathcal{A}, \delta)$ will be called an (*ordinary*) *differential algebra*.

Consider a second order linear differential operator with coefficients in $\mathcal{A}$:

$$D = \delta^2 + a_1 \delta + a_0 \quad . \tag{7.1}$$

Let us now try to factor D:

$$D = (\delta + b_0)(\delta + c_0)$$

$$= \delta^2 + b_0 \delta + \delta c_0 + b_0 c_0$$

$$= \delta^2 + (b_0 + c_0) \delta + \delta(c_0) + b_0 c_0 \tag{7.2}$$

Comparing (4.1) and (4.2), we have

$$b_0 + c_0 = a_1 \tag{7.3}$$

$$\delta(c_0) + b_0 c_0 = a_0 \ .$$

This leads to a differential equation for c_0:

$$\delta(c_0) = a_0 - (a_1 - c_0) c_0 \ . \tag{7.4}$$

This is a Riccati equation for c_0.

Suppose now that (Γ, δ) is another ordinary cummutative differential algebra which contains $(\mathcal{A}, \delta)$ as a subalgebra. We will suppose also that Γ is generated (as a differential algebra) by $\mathcal{A}$, and two linearly independent (over the scalars) solutions γ_1, γ_2 of the differential equation

$$D\gamma = 0 \ . \tag{7.5}$$

Given $c_0 \in \Gamma$ satisfying (7.4), suppose there is a $\gamma_0 \in \Gamma$ such that:

$$c_0 = - \delta(\gamma_0) \gamma_0^{-1} \ . \tag{7.6}$$

Then

$$\delta^2(\gamma_0) = - \delta(\gamma_0 c_0)$$

$$= - \gamma(\delta_0) c_0 - \gamma_0 \delta(c_0)$$

$$= - \delta(\gamma_0) c_0 - \gamma_0 (a_0 - (a_1 - c_0) c_0) \ .$$

We conclude that γ_0 satisfies the differential equation $D\gamma = 0$, hence there are scalars k_1, k_2 such that:

$$\gamma_0 = k_1 \gamma_1 + k_2 \gamma_2 \tag{7.7}$$

$$c_0 = (-k_1 \delta(\gamma_1) + k_2 \delta(\gamma_2))(k_1 \gamma_1 + k_2 \gamma_2) \tag{7.8}$$

$$k_0 = a_1 - (-k_1 \delta(\gamma_1) + k_2 \delta(\gamma_2)(k_1 \gamma_1 + k_2 \gamma_2)) \tag{7.9}$$

Let $\mathscr{D}^1$ be the Lie algebra (under commutator) of all first order differential operators: $\Gamma \to \Gamma$, i.e., those of the form

$$\Delta = \alpha\delta + \beta$$

$$\alpha, \beta \in \Gamma .$$

Now, let $\mathscr{L}$ be the Lie algebra of operations on Γ generated by

$$D'' = \delta + b_0$$
$$D'' = \delta + c_0 .$$
(7.10)

Formulas (7.8) and (7.9) tell us explicitly how the Lie algebra $\mathscr{L}$ depends on the constants k_1, k_2. In fact, they are homogeneous of degree zero in k_1, k_2, hence there is a map

$$P_1(K) \to (\text{Lie subalgebra of } \mathscr{D}^1) . \quad (7.11)$$

($P_1(K)$ denotes the projective space of one-dimension based in the field K.) Thus, we have a "deformation" of Lie subalgebras of $\mathscr{D}^1$.

8. DIFFERENTIAL EQUATIONS WHOSE LIE ALGEBRAS ARE FINITE DIMENSIONAL

Continue with $\mathscr{A}$, δ, D as in Section 7. We have assigned to D a one-parameter family of Lie algebras. An obvious question suggests itself. When are these Lie algebras finite dimensional?

Suppose then that $\mathscr{L}$ is generated by operations

$$D' = \delta + b_0$$

$$D'' = \delta + c_0$$

$$[D', D''] = \delta(c_0) - \delta(b_0)$$

$$[D', [D', D'']] = \delta^2(c_0) - \delta^2(b_0)$$

$$\vdots$$

Thus, we see that $\mathscr{L}$ is finite dimensional if and only if there exist scalars $\sigma_0, \sigma_1, \ldots, \sigma_{n-1}$ such that:

$$\sigma(c_0 - b_0) + \sigma_1 \delta(c_0 - b_0) + \cdots + \delta^n(c_0 - b_0) = 0 \ . \tag{8.1}$$

This means that $c_0 - b_0$ satisfies a linear differential equation with *constant coefficients*.

$$\Delta(c_0 - b_0) = 0 \ . \tag{8.2}$$

Thus, the coefficients $\sigma_0, \sigma_1, \ldots, \sigma_{n-1}$ of Δ (where Δ is chosen to be the least order constant coefficients operator annihilating $c_0 - b_0$) are the *invariants* of the Lie algebra.

The condition that the Lie algebra $\mathscr{L}$ be finite dimensional can then be turned into conditions in the coefficients a_0, a_1 of the differential operator D by means of Equations (8.3)-(8.4).

$$\delta(c_0 - b_0) = a_0 - b_0 c_0 - \delta(a_1) - \delta(c_0)$$

$$= -b_0 c_0 + (a_1 - c_0)c_0 - \delta(a_1)$$

$$\delta^2(c_0 - b_0) = -\delta(b_0)c_0 - b_0 \delta(c_0) + (\delta(a_1) - \delta(c_0))c_0$$

$$+ (a_1 - c_0)\delta c_0 - \delta^2(a_1) \ .$$

Using (8.4) again, $\delta^2(c_0 - b_0)$ can be expressed in terms of expressions where derivatives of c_0, b_0 do not occur. Thus, we see that ultimately, relation (7.1) is expressed via a polynomial equation connecting a_0, b_0, c_0, and c_1. It does not seem practical to solve these equations explicitly.

9. GENERALIZATION OF THE FOCK-BARGMANN-SEGAL CONSTRUCTION

Let us now turn this construction around. Start off with the differential algebra $(\mathscr{A}, \delta)$, and to two elements $b_0, c_0 \in \mathscr{A}$. Construct the Lie algebra $\mathscr{L}$, as the Lie algebra of linear maps $\mathscr{A} \to \mathscr{A}$ generated (under commutation) by the elements

Quantum Stochastic Systems

$$D' = \delta + b_0$$

$$D'' = \delta + c_0 \ .$$

Thus,

$$[D',D''] = \delta(c_0 - b_0)$$

$$[D'',[D',D'']] = [D',[D',D'']]$$

$$= \delta^2(c_0 - b_0) \ ,$$

etc.

We can think of this abstractly as giving a Lie algebra $\mathscr{L}$, and a representation of $\mathscr{L}$ into the Lie algebra of differential operators of the differential algebra $(\mathscr{A},\delta)$, which we denote as $\mathscr{D}$.

However, other representations are possible for the Lie algebra $\mathscr{L}$. Geometrically, this gives rise to representations of $\mathscr{L}$ in terms of *pseudo-differential operators*. Let us formulate abstractly what is involved here.

Consider a Lie algebra $\mathscr{G}$. Construct the associative universal algebra $U(\mathscr{G})$. $U(\mathscr{G})$ consists of the polynomials

$$A_1 \cdots A_n$$

in the elements of $\mathscr{G}$, subject only to the associative law and to the relations:

$$[A_1,A_2] = A_1 A_2 - A_2 A_1 \ .$$

Gelfand and Kirillov then construct [12] the "quotient field" of "rational functions",

$$A_1^{-1} A_2 \cdots A_n^{-1} \cdots$$

and denote this as $QU(\mathscr{G})$. It too is a Lie algebra.

Suppose that we have another representation

$$\rho': \mathscr{L} \to QU(\mathscr{G})$$

for some Lie algebra $\mathcal{G}$. We might be able to construct a representation

$$\alpha: \mathcal{G} \to \mathcal{G} .$$

α then extends to $U(\mathcal{G})$.

Let us suppose that the algebra $\mathcal{D}$ of differential operators on $\mathcal{A}$ can be enlarged to another Lie algebra $\mathcal{PD}$ of operators called *pseudo-differential operators*. Let us also say that α can be extended to a representation

$$\alpha: AU(\mathcal{G}) \to \mathcal{PD} .$$

Suppose finally that there is given another Lie algebra representation

$$\rho': \mathcal{L} \to QU(\mathcal{G}) .$$

Compose this with α_1 to obtain a Lie algebra homomorphism

$$\alpha\rho': \mathcal{L} \to \mathcal{PD} .$$

Thus, we have two representations of $\mathcal{L}$ into the "pseudo-differential operators": the original representation ρ into the *differential* operators and the $\alpha\rho'$ we have just constructed. *It then may be possible to construct a linear map*

$$\beta: \mathcal{A} \to \mathcal{A}$$

which intertwines these two representations of $\mathcal{L}$.

I will now show that the Fock-Bargmann-Segal construction, which is so important in quantum mechanics, is another special case. In fact, in this case, $\rho'(\mathcal{L})$ actually lies in $\mathcal{G}$, i.e., there is nothing "pseudo". However, this is probably a very "degenerate" situation that happens rarely.

Let $\mathcal{A}$ be the complex-valued functions which are analytic in the complex variable z. Let

$$\delta = \frac{d}{dz}$$

$$D' = \delta + \frac{z}{2}$$

$$D'' = \delta - \frac{z}{2}$$

$$[D', D''] = 1 \quad .$$

Another representation is

$$\rho'(D') = z'$$

$$\rho'(D'') = \delta' \quad .$$

If one now looks for integral operators

$$\beta(f) = \int k(z, z') f(z') \, dz' \quad ,$$

which intertwines the two representations, one comes back to the usual formulas.

10. AN ABSTRACT FORM OF PICARD-VESSIOT THEORY

We have seen that the theory of factorization of second order linear differential operators is an important topic, linking the mathematics and physics of stochastic systems. Thus, it pays to take a look at the cla-sical Picard-Vessiot theory, which deals precisely with such factorization problem. We shall now present a development of an algebraic form of the theory.

Let Γ be a vector space over a scalar field. (The scalar field should have zero characteristic. In the applications it will be the real or complex numbers.) Let $L(\Gamma)$ denote the algebra of linear maps: $\Gamma \to \Gamma$. Let $GL(\Gamma)$ denote the group of automorphisms of Γ.

An *algebraic model of a differential structure* on Γ will be defined by a linear map

$$\delta : \Gamma \to \Gamma \quad ,$$

and an *algebra* (under composition) $\mathscr{A}$ of linear operators on Γ such that the following condition is satisfied:

$$[a,\delta] \equiv a\delta - \delta a \in \mathscr{A} \tag{10.1}$$

for all $a \in \mathscr{A}$.

An *automorphism of the differential structure* will be an element $g \in GL(V)$ such that

$$g\delta = \delta g \tag{10.2}$$

$$gag^{-1} \in \mathscr{A} \, , \quad \text{for all } a \in \mathscr{A} .$$

A *differential operator associated with the differential structure* will be an element

$$D \in L(\Gamma)$$

which can be written in the form

$$D = a_n \delta^n + \cdots + a_0 \, , \tag{10.3}$$

where $a_0, \ldots, a_n \in \mathscr{A}$.

The smallest integer n such that $a_n \neq 0$ is called the *order* of the differential operator. Let $\mathscr{S}(D)$ be the space of solutions γ of the equation

$$D\gamma = 0 \, ,$$

i.e., algebraically, $\mathscr{S}(D)$ is the kernel of D. An automorphism g of the differential structure is said to be a *symmetry* of the differential operator if it satisfies the following condition:

$$g(\mathscr{S}(D)) \subset \mathscr{S}(D) . \tag{10.4}$$

The set of such symmetries forms a group called the *Galois group* of D, denoted $G(D)$.

In Galois theory, one is interested in two "functors". The assignment

$$D \to \mathscr{S}(D)$$

and the assignment

$$D \to G(D) \ .$$

Notice that $G(D)$ by its very nature admits a representation by linear maps in $\mathscr{S}(D)$, called the *Picard-Vessiot representation*.

This is, of course, just a collection of definitions. What is needed to give it more kick is some hypothesis about Γ and δ, and the class of D's which assume that $G(D)$ is a Lie group, and that relates the structure of $G(D)$ to the algebraic structure of the D's.

The traditional example where these conditions are satisfied is that where Γ is the space of functions of a complex variable z, which are analytic in some region of the complex z-plane $\delta = d/dz$, and the algebra $\mathscr{A}$ consists of operators of multiplication by analytic functions. The operators of form (10.3) are then the ordinary linear differential operators in the usual sense. One of the main theorems proved by Picard and Vessiot is that reducibility of the Galois group of D, as a group of linear transformations on $\mathscr{S}(D)$, is *equivalent* to reducibility of D, in the sense that it can be factored to the product of lower order operators. I will not attempt to prove this result here. Instead, let us examine it for the special case of second order ordinary linear differential operators with rational coefficients.

11. FACTORING SECOND ORDER DIFFERENTIAL OPERATORS WITH RATIONAL COEFFICIENTS

Let Γ be a vector space of analytic functions in some region of the complex plane, with coordinate z. Suppose Γ is preserved under multiplication by rational functions and differentiation. Let $\mathscr{A}$ be the algebra of linear operators on Γ resulting from multiplication by rational functions. Set

$$\delta = \frac{d}{dz} .$$

Let

$$D = \delta^2 + a_1 \delta + a_0 \qquad (11.1)$$

be a second order linear differential operator with rational coefficients a_0, a_1.

Let us now try to factor D:

$$D = (\delta + b_0)(\delta + c_0)$$

$$= \delta^2 + b_0 \delta + \delta c_0 + b_0 c_0$$

$$= \delta^2 + (b_0 + c_0)\delta + \delta(c_0) + b_0 c_0 . \qquad (11.2)$$

Composing (11.1) and (11.2), we have

$$b_0 + c_0 = a_1$$

$$\delta(c_0) + b_0 c_0 = a_0 . \qquad (11.3)$$

This leads to a differential equation for c_0:

$$\delta(c_0) = \frac{dc_0}{dz}$$

$$= a_0 - (a_1 - c_0)c_0 . \qquad (11.4)$$

This is a Riccati equation for c_0, which is well known to be as difficult to solve as the Equation (11.1).

To deal with this in the fashion of contemporary differential algebra, let $\mathscr{F}$ be the smallest field of functions containing the rational functions and the two linearly independent solutions of $D\gamma = 0$. The Galois group of

Quantum Stochastic Systems

D is then defined as the subgroup of the automorphism of $(\mathscr{F}, \delta)$ which leave invariant the rational functions.

Now, a solution of the Riccati equation can be obtained from a particular solution y of $Dy = 0$. Set:

$$c = -\delta(y) y^{-1} . \tag{11.5}$$

Then,

$$\begin{aligned} \delta(c) &= -\delta^2(y) y^{-1} + \delta(y)^2 y^{-2} \\ &= (\delta(y) a_1 + a_0 y) y^{-1} + c^2 \\ &= -c a_1 + a_0 + c^2 . \end{aligned} \tag{11.6}$$

Thus, c and c_0 satisfy the same Riccati differential equation. The steps are reversible; if c_0 is a solution of (11.4), then y_0 such that

$$-(y_0) \cdot y_0^{-1} = c_0 \tag{11.7}$$

is a solution of $Dy_0 = 0$.

Let y_1, y_2 be any two linearly independent solutions of $Dy = 0$, and let y_0 be defined by (11.7). Then, y_0 is a solution of $Dy = 0$. Hence:

There are constants λ_1, λ_2 such that:

$$y_0 = \lambda_1 y_1 + \lambda_2 y_2 . \tag{11.8}$$

Now, if D factors in the form (11.2) *with rational coefficients*, then there is a c_0, a rational function, and λ_1, λ_2 complex numbers, such that (11.7) and (11.8) hold.

Now, suppose g is an element of the Galois group of D's, i.e., g is an automorphism of the differential field $\mathscr{F}$ which leaves invariant the rational functions. Then, g maps solutions of $Dy = 0$ into solutions

$$g(y_1) = \sigma_{11} y_1 + \sigma_{12} y_2$$

$$g(y_2) = \sigma_{21} y_1 + \sigma_{22} y_2 \;,$$

where $\sigma_{11}, \ldots, \sigma_{22}$ are scalars. Also,

$$g(c_0) = c_0 \;,$$

since c_0 is rational. Since g commutes with $\delta = d/dz$, we see that:

g leaves invariant the linear subsapce of $\mathscr{S}(D)$ spanned by y_0.

Thus, in this case, we see quite explicitly how reducibility of the Galois group of D is equivalent to factorizability of D in terms of differential operators *with rational coefficients*.

12. THE INFELD-HULL FACTORIZATION OF THE BESSEL EQUATION

To see a non-trivial example of factorization in the Infeld-Hull sense, take Γ to be a space of an analytic function of the complex variable z, in a region Z of the complex plane. Let $\delta = d/dz$. Set

$$D = \delta^2 + z^{-1}\delta - n^2 z^{-2} \;. \qquad (12.1)$$

Then, the eigenvectors of D with eigenvalue $\lambda = 1$ are the solutions of the Bessel equation. Set:

$$A_j = \delta + jz^{-1} \;. \qquad (12.2)$$

Let us try to write D as a product:

$$\begin{aligned} D &= A_j A_k \\ &= (\delta + jz^{-1})(\delta + kz^{-1}) \\ &= \delta^2 + (j+k)z^{-1}\delta - kz^{-2} + jkz^{-2} \;. \qquad (12.3) \end{aligned}$$

Compare (12.3) and (12.1):

$$j + k = 1$$

$$-n^2 = jk - n$$

$$= j(1-j) - n$$

$$-n^2 + n = j - j^2$$

or

$$j = n, \quad k = 1 - n$$

$$D = (\delta + nz^{-1})(\delta - (n-1)z^{-1}) \tag{12.4}$$

$$= A_n A_{-n+1} . \tag{12.5}$$

The Lie algebra generated by the differential operators A is an interesting one.

$$[A_n, A_n] = [\delta + nz^{-1}, \delta + mz^{-1}]$$

$$= (n - m)B_2 , \tag{12.6}$$

with

$$B_2 = z^{-2} .$$

Set:

$$B_n = z^{-n} .$$

Then,

$$[A_n, B_m] = B_{m+1} \tag{12.7}$$

$$[B_n, B_m] = 0 . \tag{12.8}$$

Let $\mathscr{L}$ be the Lie algebra generated by the vector fields A_n. It has an interesting graded structure:

$\mathscr{L}^1$ = subspace spanned by the A_n

$\mathscr{L}^{1+j}$ = one-dimensional subspace spanned by B_{1+j}, $j = 1, 2, \ldots$

Then,

$$[\mathscr{L}^j, \mathscr{L}^k] \subset \mathscr{L}^{j+k} .$$

Set:

$$\mathscr{L}_j = \mathscr{L}^j + \mathscr{L}^{j+1} + \cdots .$$

The $\mathscr{L}_j$ determine a filtered Lie algebra. Then, $\mathscr{L}_2, \mathscr{L}_3, \ldots$ form abelian ideals. This exhibits a sort of "generalized nilpotent" structure of $\mathscr{L}$.

13. THE INFELD-HULL FACTORIZATION OF THE BESSEL EQUATION IN TERMS OF DEFORMATION THEORY

What we have done is to find a one-parameter family of differential operators

$$\lambda \to D_\lambda$$

such that:

For $\lambda = 1$, D is the Bessel operator itself.

For $\lambda = 0$, D factors over the rationals.

Thus, the Galois group

$$\lambda \to G_\lambda(D_\lambda)$$

is a one-parameter family of groups—a "deformation"—which becomes reducible for $\lambda = 0$.

14. THE INFELD-HULL FACTORIZATION FOR THE WHITTAKER EQUATION

The Whittaker functions are generalizations of the Bessel functions. (From the Lie group point of view, both sorts of functions are matrix elements of representations of solvable Lie groups.)

Consider the following Lie algebra of first order differential operators generated by the vector field

$$V = z^{1/2} \frac{d}{dz}$$

and the functions

$$f_1 = z^{1/2}$$

$$f_2 = z^{-1/2} \ .$$

According to [25], page 415, we have:

$$\left(V - \mu f_2 + \frac{1}{2} f_1\right)\left(V + \frac{2\mu - 1}{2} f_2 - \frac{1}{2} f_1\right) - \left(\mu - \lambda - \frac{1}{2}\right)$$

$$= z^{-1}\left(\frac{d^2}{dz^2} - \frac{1}{4} + \lambda z^{-1} + \left(\frac{1}{4} - \mu^2\right) z^{-2}\right)$$

This is, of course, an Infeld-Hull type of factorization.

It is interesting to calculate the Lie algebra $\mathscr{L}$ generated by the V, f_1, f_2. (The expression on the left hand side of (14.1) is then on the universal enveloping algebra of $\mathscr{L}$.)

$$V(f_1) = \frac{1}{2}$$

$$V(f_2) = -\frac{1}{2} z^{-1}$$

$$V(z^{-1}) = -z^{1/2} z^{-2}$$

$$= -z^{-3/2}$$

and so on .

Notice that the functions obtained in this way are always rational functions in $z^{1/2}$. Thus, $\mathscr{L}$ can be described very elegantly in terms of Riemann surface theory.

Theorem 14.1. The Lie algebra $\mathscr{L}$ involved in the recursion relations and the Infeld-Hull structure of the Whittaker functions can be realized as meromorphic, first order differential operators on the Riemann surface of the algebraic curve

$$w^2 = z .$$

15. THE LIE ALGEBRA AND INFELD-HULL STRUCTURE FOR THE LEGENDRE FUNCTION

The process we have gone through for Bessel's equation and the Whittaker functions can be repeated for virtually every "special function" given in the treatise by Vikenkin. For example, on page 137, he proves the following formula:

$$(V - f_{n+1,m})(V + f_{n,m}) - (\ell - n)(\ell + n + 1)$$

$$= (1 - z^2) \frac{d^2}{dz^2} - 2z \frac{d}{dz} + \frac{m^2 + n^2 - 2mnz}{1 - z^2} + \ell(\ell + 1) \tag{15.1}$$

where

$$V = \sqrt{1 - z^2} \; \frac{d}{dz} \tag{15.2}$$

$$f_{n,m} = \frac{nz - m}{\sqrt{1 - z^2}} \tag{15.3}$$

Let $\mathscr{R}$ be the field of rational functions on the Riemann surface of the algebraic curve

$$w^2 + z^2 = 1 . \tag{15.4}$$

Quantum Stochastic Systems

V is a derivation of this field, and each $f_{n,m}$ is an element of $\mathcal{R}$. Thus, the Lie algebra $\mathcal{L}$ generated by the operators appearing in the Infeld-Hull factorization of the operator on the right hand side of (15.4) is a Lie algebra of first order rational differential operators on the Riemann surface.

16. THE LIE ALGEBRA OF THE INFELD-HULL RELATIONS IN TERMS OF THE RATIONAL ENVELOPING ALGEBRAS OF FINITE DIMENSIONAL LIE ALGEBRAS

In previous sections we have presented some calculations which seem to indicate that the Lie algebra $\mathcal{L}$ which generates certain of the Infeld-Hull factorizations are naturally interpreted in terms of compact Riemann surfaces. However, the functions we have been dealing with appear as matrix elements of representations of Lie groups. Thus, we might also look for some more direct relation to Lie groups. We will restrict attention here to the Bessel equation.

Let M be the unit circle in R^2, parameterized by real θ, $0 \leq \theta < 2\pi$. Let $\mathcal{F}(M)$ be the C^∞, complex-valued functions on M. Let Z denote the complex numbers, with complex coordinate z. Let $\mathcal{A}(Z)$ be the space of analytic functions on Z. Let

$$B: \mathcal{F}(M) \to \mathcal{F}(Z)$$

be the following mapping:

$$B(f)(z) = \int_0^{2\pi} e^{iz \sin \theta} f(\theta) \, d\theta \quad . \tag{16.1}$$

Let $\mathcal{G}$ be the Lie algebra of operations on $\mathcal{F}(M)$ spanned by the following three:

$$C_1 = \frac{\partial}{\partial \theta}$$

$$C_2 = \sin \theta$$

$$C_3 = \cos \theta \quad .$$

We see that $\mathcal{G}$ is a solvable Lie algebra, isomorphic to the group of rigid motions in $\mathbb{R}^2$. $\mathcal{L}$, the Lie algebra which is generated by the Infeld-Hull factorization for the Bessel functions, is generated by the operators

$$\frac{d}{dz}, \quad z^{-1}.$$

Now,

$$\frac{d}{dz}(B(f)) = iB(C_2(f)).$$

Set:

$$C_4 = -i(\cos\theta)^{-1}\frac{d}{d\theta}$$

$$= -iC_3^{-1}C_1.$$

Then,

$$z(B(f)) = B(C_4(f)).$$

Thus, at least formally,

$$z^{-1}B(f) = B(C_4^{-1}(f))$$

$$= B(iC_1^{-1}C_3).$$

Thus, the Lie algebra $\mathcal{L}$ is essentially generated by

$$A_i C_2, \quad iC_1^{-1}C_3.$$

Let us check this out with the commutation relations:

$$[C_1, C_2] = C_3$$

$$[C_1, C_3] = -C_2$$

$$[C_2, C_3] = 0.$$

Thus,

Quantum Stochastic Systems

$$[C_2, C_1^{-1} C_3] = [C_2, C_1^{-1}] C_3$$

$$= - C_1^{-1} [C_2, C_1] C_1^{-1} C_3$$

$$= (C_1^{-1} C_3)^2 \quad . \tag{16.2}$$

Again, we see that C_2 and $C_1^{-1} C_3$ generate a graded Lie algebra of a generalized nilpotent type.

Now, algebraically, objects like $C_1^{-1} C_3$ are elements of the "field of fractions of the universal enveloping algebra of $\mathscr{L}$", denoted as $\mathscr{F}(\mathscr{L})$. Analytically, they are "pseudo-differential operators on the manifold M."

17. QUANTUM STOCHASTIC SYSTEMS WHOSE DETERMINISTIC PART IS THE HARMONIC OSCILLATOR IN THE FOCK-BARGMAN-SEGAL REPRESENTATION

We have developed the quantum FP equations for systems whose deterministic past is the harmonic oscillator *in the traditional Schrodinger representation*. We shall see that the description is much simpler in terms of the well known *Fock-Bargman-Segal* representation.

To construct it, let H consist of the complex analytic function

$$z \to \psi(z)$$

of the complex variable $z \in \mathbb{C}$ which satisfy:

$$\int |\psi(z)|^2 e^{-|z|^2} \, dz \, d\bar{z} < \infty \tag{17.1}$$

Make H into a Hilbert space, with the following inner product:

$$\langle \psi_1, \psi_2 \rangle = \int \bar{\psi}_1^*(z) \psi_2(z) e^{-|z|^2} \, dz \, d\bar{z} \tag{17.2}$$

Let us compute the adjoint of the operator ∂_z ($\equiv \partial/\partial z$) with respect to this inner product:

$$\langle \psi_1, \partial_z \psi_2 \rangle = \int \bar{\psi}_1 \partial_z \psi_2 e^{-z\bar{z}}$$

$$= -\int \overline{\partial_{\bar{z}} \psi_1} \psi_2 e^{-z\bar{z}} - \int \bar{\psi}_1 \psi_2 (-\bar{z} e^{-z\bar{z}})$$

$$= \text{, since } \partial_{\bar{z}} \psi_1 = 0 \text{ ,}$$

$$= \int \overline{z\psi_1} \psi_2 e^{-z\bar{z}} \, dz \, d\bar{z}$$

$$= \langle z\psi_1, \psi_2 \rangle \qquad (17.3)$$

We deduce:

> The adjoint of the operator ∂_z is the operator z (i.e., multiplication by z). (17.4)

Thus, ∂_z and z are operators which are adjoints of each other and satisfy the Heisenberg commutation relation:

$$[\partial_z, z] = 1 \text{ .}$$

Let

$$h = z\partial_z \text{ .}$$

The adjoint of h is:

$$h^* = \partial_z^* z^*$$

$$= z\partial_z \text{ ,} \qquad (17.5)$$

i.e., h^* is Hermitian. It is the Hamiltonian for the harmonic oscillators (in the Fock-Bargamnn-Segal representation).

Thus, we might set up the quantum FP-equation based on this h as the deterministic part:

$$\frac{\partial \rho}{\partial \tau} = [h, \rho] + \sum_i A_i \rho B_i \text{ ,} \qquad (17.6)$$

Quantum Stochastic Systems

where ρ is a Hermitian operator: $H \to H$ of trace one. The simplest and most obvious choice (based on the Lie theoretic principle discussed earlier) for the A's and B's is that they should satisfy:

$$[h,A] = A . \tag{17.7}$$

One obvious choice satisfying these commutation relations is:

$$A = \partial_z \text{ or } z . \tag{17.8}$$

Let us then work out what seems to be the simplest choice for the FP-equation (17.6):

$$\frac{\partial \rho}{\partial \tau} = [h,\rho] + A\rho B \tag{17.9}$$

$$A = \partial_z , \quad B = z .$$

(Since the adjoint of ∂_z is z, the second term on the right hand side is chosen so as to preserve the Hermitian-ness.)

Let us work out the second term on the right hand side more explicitly. Suppose that $\rho(t)$ is an integral operator with kernel $\alpha(z,y,t)$:

$$\rho(\psi)(z) = \int \alpha(z,y,t) \psi(y) e^{-|y|^2} \, dy \, d\bar{y} .$$

Then,

$$(A\rho B)(\psi)(z) = \int \partial_z \alpha(z,y,t) y \psi(y) e^{-|y|^2} \, dy \, d\bar{y} . \tag{17.10}$$

Thus the differential equation (17.9) is equivalent to the following equation for the kernel α:

$$\int \partial_\tau \alpha(z,y,t)\psi(y) e^{-|y|^2} \, dy \, d\bar{y} = h\rho(\psi)(z) - \rho h\psi(z) + A\rho B(\psi)(z)$$

$$= z\partial_z \int \alpha(z,y,t)\psi(y) e^{-|y|^2} \, dy \, d\bar{y}$$

$$- \int \alpha(z,y,t) y \partial_y e^{-|y|^2} \, dy \, d\bar{y}$$

$$+ \int \partial_z \alpha(z,y,t) y\psi(y) e^{-|y|^2} \, dy \, d\bar{y}$$

Now,

$$\int \alpha(z,y,t) y \partial_y \psi(y) e^{-|y|^2} \, dy \, d\bar{y} = -\int (\partial_y \alpha y\psi + \alpha\psi - \bar{y}\alpha y\psi) e^{-|y|^2} \, dy \, d\bar{y}$$

Thus, it is equivalent to the following differential equation:

$$\alpha_t = z\alpha_z - y\alpha_y - \alpha + |y|^2 \alpha \quad . \tag{17.11}$$

This is a *first order* linear partial differential equation which can be readily solved. We see another illustration of the most beautiful feature of the Fock-Bargamnn-Segal representation. It converts certain types of *quantum* "stochastic" FP-equations into equations that have a "deterministic" look. Of course, the reason for this is buried in the geometry--the representation is chosen so that the Hamiltonian becomes a *vector field* (i.e., a first order linear differential operator) instead of, as in the "Schrodinger" representation, a second order differential operator. Referring back to the Picard-Vessiot theory, we can see a glimmer of a general explanation of this phenomenon.

References

1. R. Abraham and J. Marsden, *Foundations of Mechanics*, 2nd ed., Addison-Wesley, Reading, MA.

2. L. Arnold, *Stochastic Differential Equations*, Wiley, New York, 1974.

3. V. Benes, to appear, *Stochastics*, 1980.

4. R.W. Brockett and J.M.C. Clark, The geometry of the conditional density equation, in *Analysis and Optimization of Stochastic Systems*, O.L.R. Jacobs (ed.), Academic Press, 1980.

5. W. Fleming and R. Rishel, *Deterministic and Stochastic Optimal Control*, Springer-Verlag, Berlin, 1975 (page 109).

6. V. Guillemin and S. Sternberg, *Geometric Asymptotics*, AMS, Providence, R.I., 1977.

7. R. Hermann and A.J. Krener, Nonlinear observability and controllability, *IEEE Trans.*, AC-22, 1977 (pages 728-740).

8. R.E. Kalman, A new approach to linear filtering and prediction problems, *Trans. ASME J. Basic Engr.* $\underline{82}$, 34-45 (1960).

9. S.K. Mitter, On the analogy between mathematical problems of nonlinear filtering and quantum physics, to appear in *Ricerche di Automatica*, Special Issue on System Theory and Physics.

10. R.E. Mortensen, Maximum likelyhood recursive nonlinear filtering, *J. Optimization Theory and App.* $\underline{2}$, 386-394 (1968).

11. B. Simon, *Functional Integration and Quantum Physics*, Academic Press, New York, 1979.

12. H. Sussmann, On the gap between deterministic and stochastic ODE's, *Ann. Prob.* $\underline{6}$, 19-41 (1978).

13. E. Wong, *Stochastic Processes in Information and Dynamical Systems*, McGraw-Hill, 1970.

14. M. Zakai, On the optimal filtering of diffusion processes, *Z. Wahr. Verw. Gebiete* $\underline{11}$, 230-243 (1969).

15. M. Lax, Quantum theory of noise in masers and lasers, in *Dynamical Processes in Solid State Optics*, 1966 Tokyo Summer Lectures in Theoretical Physics, R. Kubo and H. Kaminsura (eds.), W.A. Benjamin, New York, 1967.

16. R. Hermann, *Cartanian Geometry, Nonlinear Waves and Control Theory*, Part II (Interdisciplinary Mathematics, Volume 21), Math Sci Press, Brookline, MA, 1980.

17. R. Hermann, Infeld-Hull factorization and Galois-Picard-Vessiot theory for differential operators, preprint.

18. R. Hermann, Infeld-Hull factorization and the filtering problem, preprint.

19. R. Hermann, *Geometry, Physics and Systems*, Marcel Dekker, N.Y., 1972.

20. E. Picard, *Traité d'Analyse*, Vol. 3, Chapter 7, Gauthier-Villars, Paris, 1928.

21. E. Vessiot, Sur les integrations des équations différentielles lineaires, *Am. Sci. Ecole Norm.* **131**, 9 (1892), 192-280.

22. I. Kaplansky, *Introduction to Differential Algebra*, Herman, Paris, 1957.

23. F. Kolchin, *Differential Algebra and Algebraic Groups*, Academic Press, 1973.

24. L. Infeld and T.E. Hull, *Rev. Mod. Phys.* **123**, 21 (1951).

25. N.J. Vilenkin, *Special Functions and Group Representations*, American Mathematical Society, 1968.

26. I.M. Gelfand and A.A. Kirillov, The structure of the Lie field connected with a semi-simple Lie algebra, *Func. Analy. Appl.* **3**, 6 (1969).

27. R.W. Brockett, On the invariance group of the conditional density equations, *Proc. IEEE Decision and Control Conf., 1979*, Ft. Lauderdale, Florida.

28. R.W. Brockett, Remarks on finite dimensional nonlinear estimation, *Conf. on Algebraic and Geometric Methods in System Theory*, Bordeaux, France, September 1978.

29. S. Mitter, Filtering theory and quantum fields, *Conference on Algebraic and Geometric Methods in System Theory*, Bordeaux, France, September 1978. To appear in *Asterisque*, 1980.

30. J. Dixmier, *Algèbres Envelloppantes*, Gauthier-Villars, Paris, 1974.

31. I.E. Segal, Tensor algebras over Hilbert spaces, I, *Trans. Am. Math. Soc.* **81**, 106-134 (1956).

32. M. Fujisaki, G. Kallianpur, and H. Kunita, Stochastic differential equations for the non-linear filtering problem, *Osaka J. Math.* **9**, 19-40 (1972).

33. R.E. Mortensen, Optimal control of continuous time stochastic systems, Doctoral Dissertatation, Dept. of EE, Univ. of California, Berkeley, 1966.

34. L. Schlesinger, *Hanbuch der Theories der Lineardifferentialgleuchungen*, Teubner, Leipzig, 1897.

35. H. Weyl, *The Concept of a Riemann Surface*, Addison-Wesley, 1964.

36. P. Griffiths and J. Harris, *Principles of Algebraic Geometry*, Wiley, 1978.

37. N. Hurt and R. Hermann, *Quantum Statistical Mechanics and Lie Group Harmonic Analysis*, Math Sci Press, 1980.

38. O. Hijab, "Minimal Energy Estimation", thesis, University of California, 1980.

39. M. Davis and S. Marcus, An introduction to nonlinear filtering (to appear).

40. M. Hazewinkel and S. Marcus, On Lie algebras and finite dimensional filtering (to appear).

Chapter 14

LIE THEORY AND STOCHASTIC SYSTEMS, PART II
THE GEOMETRY OF A PROBABILISTIC LIE THEORY

1. INTRODUCTION

It has long been a goal of system theorists to build a Lie-theoretic bridge between stochastic and deterministic system theory. I differ somewhat from others in that the stochastic differential equation-Ito calculus does not appeal to me aethetically. Instead, I would like to construct a more direct geometric link between the deterministic and stochastic theory. In this paper I will explore, intuitively, using the concepts of modern differential geometry, several ways that this might be done. I have especially in mind developing certain links between such a stochastic differential geometry and a related approach to the geometry of quantum mechanics I have developed

2. TANGENT VECTORS AND VECTOR FIELDS FOR THE SPACE OF PROBABILITY MEASURES

In order to make a "Lie theory" for spaces of probability measures, we first have to define *target vectors* and *vector fields*. I will do this by following certain differential-geometric intuition.

Remark. I have suggested an approach to quantum mechanics which replaces the standard Hilbert space with the set of ordered pairs

$$(D(q)dq, \omega) ,$$

where q is configuration space of the mechanical system, $D(q)dq$ is a probability measure and configuration space, and ω is a closed one-form and configuration space. This space has a natural *symplectic structure*.

Let X be an orientable manifold of dimension n, with a fixed volume element differential form dx. Let $\mathcal{D}^n(X)$ be the space of smooth n-th order differential forms. Then an $\omega \in \mathcal{D}^n(X)$ is of the form:

$$\omega = f(x) dx$$

with $f \in \mathcal{F}(X)$ a C^∞, real-valued function on X.

Such a form is called a *probability form* if:

$$f(x) \geq 0$$

and

and

$$\int f(x)\, dx = 1 \quad . \tag{2.1}$$

Let $\mathscr{P}(X)$ or $\mathscr{P}$ denote the space of such forms satisfying (2.1).

Remark. A $\omega \in \mathscr{P}$ defines a probability measure on X, in the usual sense. The measure of a set is the integral

$$\int f\omega \quad ,$$

which is the characteristic function of that set. Of course, this only captures the smooth measures. To describe others in the same way, e.g., the Dirac delta function, one must pass from the smooth forms to what de Rham called *currents*.

$\mathscr{P}(X)$ is the basic object we want to study. Geometrically, it is a *hyperplane* in a linear space, so its tangent space is obvious.

Definition. The *tangent space* to a $\omega \in \mathscr{P}$ is the set of all $\theta \in \mathscr{D}^n(X)$ such that

$$\int_X \theta = 0 \quad . \tag{2.2}$$

A *vector field* to $\mathscr{P}$ is a map

$$V: \mathscr{P} \to \mathscr{D}^n(X)$$

such that $V(\omega)$ satisfies (2.2) for all $\omega \in \mathscr{P}$.

Denote the set of all maps V satisfying (2.2) by

$$\mathscr{V}(\mathscr{P}) \quad .$$

Such a vector field is said to be *linear* if it is the restriction to $\mathscr{P}$ of a linear map

$$\mathscr{D}^n(X) \to \mathscr{D}^n(X) \quad .$$

Denote the set of such linear vector fields by

$$L\mathscr{V}(\mathscr{P}) \quad .$$

Just as for the finite dimensional situation, the natural Lie algebra structure on vector fields reduces to the *commutator* on $L\mathcal{V}(\mathcal{P})$.

Examples.

a) *Vector field on probability measures defined by vector fields on* X. *Drift or Liouville operator.*

Let $\mathcal{V}(X)$ denote the usual vector fields on the manifold X; algebraically they are the derivations of the algebra of real-valued functions on X. For $V \in \mathcal{V}(X)$ let $\theta(V)$ be the linear map

$$\mathcal{D}^n(X) \to \mathcal{D}^n(X)$$

defined by Lie derivation:

$$\theta(V)(\omega) = \mathcal{L}_V(\omega) \quad . \tag{2.3}$$

To verify that $\theta(V) \in L\mathcal{V}(\mathcal{P})$, note that:

$$\int_X \theta(V)\omega = \int_X \mathcal{L}_V(\omega)$$

$$= \int_X V \lrcorner \, d\omega + \int_X d(V \lrcorner \, \omega) \quad .$$

Now, $d\omega = 0$, since ω is a top-degree form; the second term also vanishes by Stoke's theorem. (Of course, we must specify ω more precisely for this to happen. For simplicity, suppose we are dealing (if X is noncompact) with a space of forms ω which are "rapidly decreasing" at infinity.) In probability theory, these operators are called *drift operators*. In physics they are called *Liouville operators*.

b) *Hodge operators.*

Let $\delta: \mathcal{D}^n(X) \to \mathcal{D}^{n-1}(X)$

be a linear idfferential operator. (In the original Hodge theory δ is the dual with respect to a Riemannian metric of the exterior derivative.) Let

$$\theta(\delta) = d\delta \quad .$$

Again, Stokes' formula guarantees that $\theta(\delta)$ belong to $L\mathcal{V}(\mathcal{P})$. $\theta(\delta)$ is a second order differential operator.

c) *Fokker-Planck operator.*

The operators $\Delta \in L\mathscr{V}(\mathscr{P})$ of the form

$$\theta(V) + \theta(\delta)$$

may be given.

3. GENERAL FOKKER-PLANCK OPERATOR

We can unify and generalize the two examples of the last section by defining a general Fokker-Planck (F-P) operator on a manifold X as a linear operator

$$\Delta : \mathscr{D}^n(X) \to \mathscr{D}^n(X)$$

of the form

$$\Delta = d\delta \quad,$$

where δ is a linear differential operator

$$\mathscr{D}^n(X) \to \mathscr{D}^{n-1}(X) \quad.$$

Thus, the drift operators are those where δ is a first order linear differential operator, the Hodge operators where δ is a second-order linear differential operator.

Let us do the case $X = P$ as an illustration

$$\omega = P(X) \, dx$$

$$\theta(\omega) = D(P) \quad,$$

where D is a linear differential operator. Set

$$D(P) = a_0 P + a_1 P_x + \cdots \quad.$$

Then,

$$\Delta(\omega) = ((a_0 P)_x + (a_1 P_x)_x + \cdots) \, dx$$

4. THE LIE STRUCTURE OF F-P OPERATORS

Suppose

$$\Delta = d\delta \quad,$$

$$\Delta' = d\delta' \quad,$$

Stochastic Systems 243

where δ, δ' are linear maps: $\mathcal{D}^n(X) \to \mathcal{D}^{n-1}(X)$. Then:

$$[\Delta, \Delta'] = d\delta d\delta' - d\delta' d\delta \qquad (4.1)$$

$$= d(\delta d\delta' = \delta' d\delta) \ .$$

The simplest case is where Δ is a Liouville operator, i.e., Lie derivative with respect to a vector field V on X

$$\Delta(\omega) = \mathscr{L}_V(\omega) \ .$$

Then,

$$[\Delta, \Delta'] = \mathscr{L}_V d\delta' \omega - d\delta' \mathscr{L}_V \omega$$

$$= d\mathscr{L}_V \delta'(\omega) - d\delta' \mathscr{L}_V(\omega) \ .$$

Set

$$\mathscr{L}_V(\delta') = \mathscr{L}_V \delta' - \delta' \mathscr{L}_V \ ,$$

the *Lie derivative* of δ'. We have proved:

<u>Theorem 4.1</u>. The commutator of a Liouville operator with vector field V and an F-P operator with Hodge operator δ is again an F-P operator whose Hodge operator is the Lie derivative of δ by V.

Here is the general case.

<u>Theorem 4.2</u>. Let Δ and Δ' be F-P operators associated with Hodge operator δ and Δ'. Then, $[\Delta, \Delta'] = \Delta''$ is associated with the Hodge operator

$$\delta'' = \delta d\delta' - \delta' d\delta \qquad (4.2)$$

$$= \delta \Delta' - \delta' \Delta$$

<u>Example</u>. $X = R$, $\delta, \delta' = $ *first order differential operators*.

$$\omega = P(x)\, dx$$

$$\delta(\omega) = a(x)P + b(x)P_x$$

$$\delta'(\omega) = a'P + b'P_x$$

$$\Delta = d\delta(\omega)$$

$$= (aP)_x \, dx + (bP_x)_x \, dx$$

$$\delta'\Delta(\omega) = \delta'((aP)_x + (bP_x)_x) + b'((aP)_x + (bP_x)_x)_x$$

Thus,

$$\delta''(\omega) = a(a'P)_x + (b'P_x)_x) + b((a'P)_x + (b'P_x)_x)_x$$

$$- a'((aP)_x + (bP_x)_x) - b'((zP)_x + (bP_x)_x)_x$$

The terms of third degree cancel out. Hence, δ'' is a second order differential operator

$$\Delta'' = d\delta'' \text{ is a third order operator }.$$

It is then interesting to inquire when Δ'' is *again* a second order operator. The second order terms of δ'' are:

$$ab'P_{xx} + ba'P_{xx} + 2bb'P_{x\,xx} - a'bP_{xx} - ba'P_{xx} - 2b'b_x P_{xx}$$

$$= 2(bb'_x - b'b_x)P_{xx}$$

<u>Theorem 4.2.</u> The commutation of two second order F-P operators is, in general, a third order operator. It is again a second order operator only if a differential equation links its highest order coefficient, namely:

$$\frac{b_x}{b} = \frac{b'_x}{b'}$$

or

$$b' = cb .$$

<u>Remark.</u> This is a special case of the "Burchrall-Chaundy" theory, which has played an important role in the theory of nonlinear waves.

5. THE FOKKER-PLANCK OPERATORS ON FILTERED LIE ALGEBRAS

Let $\mathscr{FP}$ be the Lie algebra of Fokker-Planck operators. They are of the form

$$d\delta ,$$

where δ is a Hodge operator. For each integer n, let

$$\mathcal{FP}^n$$

be those which are of order n as differential operators. Thus,

$$\mathcal{FP}^1 \subset \mathcal{FP}^2 \subset \cdots$$

i.e., this defines an *ascending filtration*. Further,

$$[\mathcal{FP}^n, \mathcal{FP}^m] \subset \mathcal{FP}^{n+m-1} \tag{5.1}$$

The filtered Lie algebras with which we are most familiar is differential geometry are filtered in a *descending* way.

6. LIE ALGEBRAS OF SECOND-DEGREE, ONE-DIMENSIONAL FOKKER-PLANCK OPERATORS

Ultimately, we shall want to discuss, in general, Lie algebras of Fokker-Planck operators. They will play the role in stochastic system theory that Lie algebras of vector fields play in deterministic system theory. The obvious place to start will be those consisting of operators of degree at most *two*. (This corresponds to stochastic systems with "white noise".) Now, suppose $\mathcal{L}$ is a *Lie subalgebra* of $\mathcal{FP}^2$. By Theorem 4.2, we see that for each $\Delta, \Delta' \in \mathcal{L}$, there is a real number $c(\Delta, \Delta')$ such that

$$\Delta - c(\Delta, \Delta')\Delta' \in \mathcal{L}^1 \ .$$

Set:

$$\mathcal{L}^1 = \mathcal{L} \cap \mathcal{FP}^1 \ .$$

$\mathcal{L}^1$ is then a Lie subalgebra of $\mathcal{L}$, consisting of Liouville operators. Further,

$$\dim(\mathcal{L}/\mathcal{L}^1) = 1 \ .$$

The structure of $\mathcal{L}$ is obviously determined by that of $\mathcal{L}^1$, and the first Lie algebra cohomology of $\mathcal{L}$ with coefficients in the one-dimensional representation of $\mathcal{L}^1$ in $\mathcal{L}/\mathcal{L}^1$. $\mathcal{L}^1$ is a Lie algebra of vector fields on R. Lie proves that it is one of three Lie algebras:

One dimensional, generated by $\frac{\partial}{\partial x}$.

Two dimensional, generated by $\frac{\partial}{\partial x}$, $x\frac{\partial}{\partial x}$.

Three dimensional, generated by $\frac{\partial}{\partial x}$, $x\frac{\partial}{\partial x}$, $x^2\frac{\partial}{\partial x}$.

In the third case, the $\mathscr{L}^1$ is semisimple, and the Lie algebra cohomology vanishes. Thus, we have:

<u>Theorem 6.1</u>. If $\mathscr{L}$ is a Lie algebra of one-dimensional Focker-Planck operators consisting of at most second degree, with $\mathscr{L}^1$ three-dimensional, then $\mathscr{L}$ consists of the operators of the form:

$$\Delta(P(x)dx) = d(V \lrcorner P(x)dx) + bP_{xx}dx ,$$

where V is a vector field in x of at most second degree. (In other words, $\mathscr{L}^1$ is the Lie algebra of $SL(2,R)$.)

The analysis of the other possibilities of $\mathscr{L}^1$ are similar. All of these cases lead to the familiar formulas for Gaussian stochastic processes and the Kalman-Bucy filter.

7. LIE ALGEBRA OF SECOND DEGREE FEKKER-PLANCK OPERATORS DETERMINED BY THE CONFORMAL STRUCTURE OF A RIEMANNIAN MANIFOLD

The viewpoint of the previous section suggests an interesting generalization.

Suppose X has a Riemannian metric ds^2. Let $\mathscr{C}$ denote the Lie algebra of *infinitesimal conformal vector fields*, i.e., the vector fields V such that

$$\mathscr{L}_V(ds^2) = f_V ds^2 .$$

Let us examine the Lie algebra properties of Fekker-Planck operators of the form:

$$\Delta = d\delta ,$$

with

$$\delta(\omega) = V \lrcorner \omega + \delta_0(\omega) ,$$

with $V \in \mathscr{C}$, δ_0 the Hodge operator associated with the Riemannian metric.

The first task is to calculate the Lie derivative of the Hodge operator. Let $<,>$ denote the inner product that the Riemannian metric defines on differential forms. (Thus, $<,>$ is really the dual co-Riemannian metric.) Now,

$$\int <\omega, d\omega'> = \int <\delta_0\omega, d\omega'>$$

Stochastic Systems   247

for differential forms of compact support. Let $V \in \mathscr{C}$. Then,

$$\mathscr{L}_V(<\omega, d\omega'>) = f_V <\omega, d\omega'> + <\mathscr{L}_V(\omega), d\omega'> + <\omega, d\mathscr{L}_V \omega'>$$

$$\mathscr{L}_V(<\partial_0 \omega, \omega'>) = f_V <\delta_0 \omega, \omega'> + <\mathscr{L}_V \delta_0 \omega, \omega'> + <\delta_0 \omega, \mathscr{L}_V \omega'>$$

Thus,

$$0 = \int <\delta_0 f_V \omega, \omega'> + <\delta_0 \mathscr{L}_V(\omega), \omega'> + <\delta_0 \omega, \mathscr{L}_V \omega'>$$

$$0 = \int <f_V \delta_0 \omega, \omega'> + <\mathscr{L}_V \delta_0 \omega, \omega'> + <\delta_0 \omega, \mathscr{L}_V \omega'>$$

We deduce that:

$$f_V \delta_0 + \mathscr{L}_V \delta_0 = \delta_0 f_V + \delta_0 \mathscr{L}_V$$

or

$$\mathscr{L}_V(\delta_0) = \delta_0 f_V - f_V \delta_0 \quad .$$

Thus, for $V, V' \in$,

$$\Delta = \mathscr{L}_V + d\delta_0 ,$$

$$\Delta' = \mathscr{L}_{V'} + d\delta_0 ,$$

we have:

$$[\Delta, \Delta'] = [\mathscr{L}_V, \mathscr{L}_{V'}] + d(\mathscr{L}_V(\delta_0) - \mathscr{L}_{V'}(\delta_0)) \quad .$$

Thus it may be possible, for certain Riemannian manifolds, that the Fokker-Planck operators constructed in this way from the conformal vector fields form a Lie algebra. I plan to pursue this at a later point. It should tie into classic work by Paul Levy on the relation between conformal transformations and Brownian motion.

8. QUANTUM MECHANICS AND COTANGENT BUNDLE OF THE SPACE OF PROBABILITY MEASURES

Since its beginning in 1900, continuing to this day, one of the most mystifying aspects of quantum mechanics is its relation to probability theory. (Recall Einstein's belief that "God does not play dice".) This is usually discussed by physicists at a mystical-philosophical level--perhaps the mathematician can suggest that the mathematical relations between the two disciplines has not been adequately investigated. Paraphrasing Brecht, I would suggest the following slogan for speculative physics:

> First comes the mathematics, then the philosophy.

In 1962 I suggested a geometric way of thinking about the role of probability in quantum mechanics. (It is elaborated in *Vector Bundles*, Volume II.) This approach is attached to what the physicists call the "hydrodynamic interpretation" of quantum mechanics developed by E. Madeling in the 1920's. (Everything was done in the 1920's, then forgotten because Pauli did not approve it.) Since it is clearly closely related to the geometry-Lie-theory-of-stochastic-processes that I am developing here, let us throw it into the stew.

Let Q be an orientable manifold of dimension n (the configuration space of a mechanical system). Let $\mathscr{P}$ be the *smooth probability densities*, i.e., the C^∞ differential forms Q of degree n which are positively oriented with respect to the given orientation of Q, such that

$$\int_Q \rho = 1 .$$

is the basic space for a *stochastic differential geometry*. (Of course, it must be "completed" along the lines of de Rham's theory of currents--to add the probability measures in the usual sense.)

The *tangent space* to a $\rho \in \mathscr{P}$ consists of the one-forms θ on Q such that

$$\int_Q \theta = 0 .$$

Denote the space of such forms as

$$\mathscr{P}' .$$

Note that the *tangent bundle* to $\mathscr{P}$ is the product

$$\mathscr{P} \times \mathscr{P}' .$$

Stochastic Systems 249

9. ESTABROOK AND HARRISON'S TREATMENT OF THE SYMMETRIES OF THE ONE-DIMENSIONAL HEAT EQUATION USING THE THEORY OF EXTERIOR DIFFERENTIAL SYSTEMS

For a differential geometer, the true "Lie theory of differential equations" lies in E. Cartan's theory of exterior differential systems. What Cartan did was to attach exterior differential systems (i.e., collections of differential forms) to differential equations, in such a way that the "symmetries" of the differential equation became the "isomorphism" of the exterior system. For some reason, this beautiful and definitive approach has not penetrated into the community of physicists, applied mathematicians, engineers, fluid dynamicists, who concern themselves with the study of "symmetries". The general exception is the work of Estabrook and Harrison (JMP, Vol. 2, 653). We shall review here one of their examples, the heat equations. (It is, of course, the "diffusion-Fokker-Planck equation" for the simplest stochastic process.)

Let (x,t) denote independent variables. We are concerned with the following differential equation:

$$P_{xx} = P_t \qquad (9.1)$$

to be solved for a function $P(x,t)$ of these variables. (Subscripts denote partial derivatives.)

As usual in the Cartan theory, this is "geometricized" by "prolongation", i.e., adding enough new variables to our original space so that the solution of the differential equations may be identified with certain solution manifolds of the exterior system.

Introduce a five-dimensional space X, with variables:

$$(x, t, P, P_x, P_t) \quad .$$

(Of course, the reader has to keep the ambiguity in the notation straight-- here P_x, P_t do not denote partial derivatives, but new variables--in fact, just the natural variables in the Ehresmann jet spaces.) Set

$$\theta_1 = dP - P_x\, dx - P_t\, dt$$
$$\theta_2 = dP_x \wedge dt - P_t\, dx \wedge dt \quad . \qquad (9.2)$$

Let $\mathcal{E}$ be the exterior differential system generated by θ_1 and θ_2, i.e., the smallest set of differential forms on X containing θ_1 and θ_2, and closed under exterior differentiation of and exterior multiplication by arbitrary forms on X.

We claim that the two-dimensional solution submanifolds of $\mathcal{E}$ on which $dx \wedge dt \neq 0$ are essentially the solution of (9.1). Suppose then that Y is such a two-dimensional manifold, and that

$$\phi: Y \to X$$

is a submanifold map such that:

$$\phi^*(\theta_1) = 0$$
$$= \phi^*(\theta_2) \tag{9.3}$$

$$\phi^*(dx \wedge dt) \neq 0 \tag{9.4}$$

Condition (9.4) guarantees that the functions $\phi^*(x)$, $\phi^*(y)$ are functionally independent on Y. Since Y is two-dimensional, they form a coordinate system about each point of Y. (Another way of putting it is to say that the map $Y \to R^2$ determined by these functions is a local diffeomorphism.) Thus, there is a function $F(\ ,\ )$ of two variables such that:

$$\phi^*(P) = F(\phi^*(x), \phi^*(y)) \ .$$

With slight "abuse of notation", this can be taken to mean that P is given as a function $P(x,y)$ on the submanifold $\phi(Y)$. On this submanifold, the assumption that $\phi^*(\theta_1) = 0$ means that

$$P_x = \frac{\partial P}{\partial x} \ , \qquad \frac{\partial P}{\partial y} = P_y \ ,$$

i.e., the "virtual" partial derivatives P_x, P_y become literal.

Let us examine the meaning of the condition (9.2):

$$0 = \phi^*(dP_x \wedge dt) - \phi^*(P_t \, dx \wedge dt)$$
$$= \phi^*(dP_x) \wedge dt - \phi^*(P_t) \, dx \wedge dt$$
$$= P_{xx} \, dx \wedge dt - P_t \, dx \wedge dt \ ,$$

or

$$P_t = P_{xx} \ ,$$

since $dx \, dt$ is nonzero.

<u>Theorem 9.1.</u> $\mathcal{E}$ is generated by θ_1, $d\theta_1$ and θ_2.

<u>Proof.</u>

$$d\theta_2 = -dP_t \wedge dx \wedge dt$$
$$= dP_t \wedge dt \wedge dx$$

Stochastic Systems

i.e., $d\theta_2$ lies in the Grassmann algebra ideal generated by θ_1, $d\theta_1$ and θ_2. This proves it.

Let us turn to the search for the infinitesimal symmetries of $\mathscr{E}$ ("isovectors", in the terminology of Estabrook and Harrison), i.e., the vector fields V on X such that:

$$\mathscr{L}_V(\mathscr{E}) \subset \mathscr{E} \quad . \tag{9.6}$$

By Theorem 9.1, V satisfies (9.5) if and only if the following relations are satisfied:

$$\mathscr{L}_V(\theta_1) = a\theta_1 \tag{9.7}$$

$$\mathscr{L}_V(\theta_2) = \alpha \wedge \theta_1 + b d\theta_1 + c\theta_2 \tag{9.8}$$

where a, b, and c are functions on X, and α is a one-form.

Let us determine the vector fields V on X, satisfying (9.6) and (9.7). Equation (9.6) says that V is an *infinitesimal contact transformation*. Set

$$f = \theta_1(V)$$

$$= V \lrcorner \theta_1 \quad .$$

f is then a σ-form, i.e., a function on X. Then,

$$df = d(V \lrcorner \theta_1)$$

$$= \mathscr{L}_V(\theta_1) - V \lrcorner d\theta_1$$

$$= \text{, using (9.7),}$$

$$a\theta_1 - V \lrcorner d\theta_1 \quad . \tag{9.9}$$

Now use condition (9.8):

$$d(V \lrcorner \theta_2) + V \lrcorner d\theta_2 = \alpha \wedge \theta_1 + b d\theta_1 + c\theta_2 \tag{9.10}$$

Use (9.5):

$$V \lrcorner d\theta_2 = -V(x) d\theta_1 + dx \wedge (V \lrcorner d\theta_1)$$

$$= \text{, using (9.9),}$$

$$(df - a\theta_1) \wedge dx - V(x) d\theta_1 \tag{9.11}$$

Substitute (9.11) into (9.10):

$$d(V \lrcorner \, \theta_2) + (df - a\theta_1) \wedge dx - V(x)d\theta_1 = \alpha \wedge \theta_1 + bd\theta_1 + c\theta_2 \quad (9.12)$$

From these differential equations for V, Estabrook and Harrison have determined (on pp. 657-658) the possibilities. They show that the set of possibilities has the following form:

$$V(t) = 2k_6 t^2 + 2k_4 t + k_1 \,,$$

$$V(x) = 2k_6 tx + k_4 x - 2k_5 t + k_2 \,,$$

$$V(P_t) = (-\tfrac{1}{2}k_6 x^2 + k_5 x - 5k_6 t + k_3 - 2k_4)P_t$$

$$+ 2(k_5 - k_6 x)y - k_6 P + P_t^0$$

$$\quad (9.13)$$

$$V(P_x) = (-\tfrac{1}{2}k_6 x^2 + k_5 x - 3k_6 t + k_3 - k_4)y$$

$$+ (k_5 - k_6 x)P + P_x^0$$

$$V(P) = (-\tfrac{1}{2}k_6 x^2 + k_5 x - k_6 t + k_3)P + P^0$$

$k_1, \ldots, k_6$ are constants. P^0 is an arbitrary function, which is a solution of

$$P_t^0 = P_{xx}^0 \,. \quad (9.14)$$

The set of all these vector fields V forms a Lie algebra $\mathcal{G}$. Let $\mathcal{G}_0$ be the set of all such V's with:

$$k_1 = 0 = \cdots = k_6 \quad (9.15)$$

and $\mathcal{G}_1$ the set of all such V's with:

$$P^0 = 0 \,. \quad (9.16)$$

Theorem 9.2. (Estabrook and Harrison). $\mathcal{G}$ is a direct sum $\mathcal{G}_0 \oplus \mathcal{G}_1$ of linear subspaces such that

$$[\mathcal{G}_1, \mathcal{G}_0] \subset \mathcal{G}_0 \,, \quad (9.17)$$

i.e., $\mathcal{G}_0$ is a Lie algebra ideal.

$$[\mathcal{G}_1, \mathcal{G}_1] \subset \mathcal{G}_1 \,, \quad (9.18)$$

Stochastic Systems

i.e., $\mathcal{G}_1$ is a Lie subalgebra

$$[\mathcal{G}_0, \mathcal{G}_0] = 0 \qquad (9.15)$$

Thus, $\mathcal{G}$, although infinite dimensional, has something like a *Levi decomposition*.

Proof. This can basically be seen geometrically. The elements of $\mathcal{G}_0$ are the infinitesimal generators of the "gauge" transformations

$$P \rightarrow P + P^0, \qquad (9.16)$$

on the space of solutions of (9.1). (Of course, the condition that (9.15) be a symmetry of (9.1), i.e., map solutions into solutions, is that P_0 itself be a solution of (9.1). As Estabrook and Harrison themselves remark, the elements of $\mathcal{G}_1$ require the action of the group of Galilean transformations on R^2 on the solutions of (9.1). The proof of Theorem 9.2 now follows standard Lie-theoretic lines.

PART V

KRON-KONDO THEORY

One of the unfortunate features of the recent revival of applications of differential geometry in its "modern" manifold-fiber bundle form is that many of the important ideas due to the tensor analysis generation have been ignored or eclipsed.

The work of Gabriel Kron and the RAAG ("Research Association of Applied Geometry"), led by K. Kondo, is especially noteworthy as an attempt to use geometry in engineering problems as Einstein had in physics. Kron hitch-hiked around the world in the 1920's with a copy of Einstein in his pocket, and returned with his vision: He would geometrize Maxwell's equations and their interaction with mechanical systems (which are power systems, the field in which Kron had earned his Ph.D.) as Einstein had geometrized gravitation. Its a good idea! The reaction of the mathematicians and the then very small mathematical engineering community was, as usual, NIH. (NIH = "Not invented here". Things have not changed all that much, unfortunately!) Only the RAAG group in Japan took up Kron's ambitious ideas; they wrote a four volume collaborative treatise which has also been ignored. (For one thing, it is very hard to find. It is also difficult to read, because the tensor analysis used was impenetrable to anyone who was not raised on the stuff.)

I found Kron's first 1934 paper the most understandable for two reasons: The underlying geometric ideas are clearest there, and the Baroque nature of the tensor analysis formalism had not yet taken over.

As the applications of geometry in physics and engineering proliferate, I want to pay homage to these pioneers, Kron, and the RAAG memoirists.

Chapter 15

THE METHODS OF KRON, HOFFMAN AND KONDO IN GEOMETRIC SYSTEM THEORY

Abstract

Before the codification of modern "system theory", Kron and Hoffman, Kondo and the RAAG Group suggested a natural way of coupling geometric structures and linear systems. This idea is developed in terms of modern fiber bundle-connection theory. A set of equations is obtained that occurs in a wide variety of engineering problems of current importance.

1. INTRODUCTION

System theory has done science and engineering a great service by codifying the notion of "input-output system", and by studying the general properties of such systems. Of course, for most of the applications, it is still only feasible to study linear systems of the form

$$\frac{dx}{dt} = Ax + Bu$$
$$y = Cx .$$

(1.1)

Now, Gabriel Kron long ago suggested [1] a "geometrization" of the equations of electric systems, which have some of the flavor in a more definitive way in later work. Banesh Hoffman developed Kron's ideas [2] in a form that was very suggestive of the recent work, emphasizing--using the ideas of tensor analysis--that one should generalize (1.1) by "replacing ordinary derivatives with covariant derivatives". The RAAG Group [3] developed the ideas further, pointing out the relevance for a wide variety of scientific and engineering disciplines. However, this work was before modern connection theory. My aim here is to present the basic idea in a more definitive form. I believe this will be useful, e.g., in extending the geometric methods developed for the study of linear systems to power systems. Thus, one might hope to more fully realize Kron's dream of studying power systems with differential geometric methods.

Another possible application is to the study of aircraft control. In fact, Brian Doolin of the Ames Research Center (NASA) suggested to me this possibility. I would like to thank him for this, and for many other helpful suggestions.

Now, in classical notation we are dealing with systems which are composites of those of the form (1.1) and the system of ordinary differential equations encountered in analytical mechanics. In classical notation, they might be taken as systems of the following form:

$$\frac{dx}{dt} + \Gamma\left(q, \frac{dq}{dt}\right) x = A\left(q, \frac{dq}{dt}\right) x + B\left(q, \frac{dq}{dt}\right) u \qquad (1.2)$$

$$y = C\left(q, \frac{dq}{dt}\right) x \qquad (1.3)$$

$$\frac{d}{dt}\left(\frac{\partial T}{\partial \dot{q}}\right) - \frac{\partial T}{\partial q} = F(q, \dot{q}, x) \ . \qquad (1.4)$$

The left hand side of (1.2) is a sort of "covariant derivative". F is a force term. u is input, y is output. (There might be additional inputs in the forces, of course.) T is a function of $q, \dot{q}$, the "kinetic energy" of the mechanical part of the system. However, we want to study these systems in the spirit of modern "global" differential geometry, hence, we will not take this traditional approach.

2. CONNECTIONS IN VECTOR BUNDLES

Let P be a differential manifold [4,5]. Let $\mathscr{F}(P)$ denote the algebra (under multiplication) of all real-valued, C^∞ functions in P. Let $\mathscr{V}(P)$ be the vector fields in P --algebraically the derivation of $\mathscr{F}(P)$. $\mathscr{V}(P)$ is a Lie algebra under commutation

$$[V_1, V_2] = V_1 V_2 - V_2 V_1$$

$$\text{for } V_1, V_2 \in \mathscr{V}(P) \ .$$

If $\phi: P \to Q$ is a differentiable map between differential manifolds, $\phi_*: T(P) \to T(Q)$ denotes the induced map on tangent vectors. $T(P)$ denotes the *tangent bundle* to P, i.e., a point of $T(P)$ is a pair (p,v), where $v \in P_p$ = space of tangent vectors to P at p.

A manifold E, together with a map $\pi: E \to P$ is a *vector bundle* with base P if the following conditions are satisfied:

a) π is a submersion map, i.e., $\pi_*(T(P)) = T(Q)$

b) Each fiber $\pi^{-1}(p)$ is a vector space. The resulting operations are differentiable.

c) The structure is local product, i.e., each point p of P has a neighborhood U, such that

$$E_U \equiv \pi^{-1}(U) \subset E$$

is isomorphic to that product $U \times R^n$ with the isomorphism preserving the linear structure in the fibers.

These properties can also be expressed algebraically in terms of the cross-section maps

$$\gamma: P \to E ,$$

i.e., those maps such that

$$\pi\gamma = \text{identity} .$$

Denote by $\Gamma(E)$ the space of cross-sections. $\Gamma(E)$ is a vector space, since the fibers are vector spaces and elements of $\Gamma(E)$ take values in these vector spaces. Further, elements of $\Gamma(E)$ can be multiplied by those of $\mathscr{F}(P)$, i.e., $\Gamma(E)$ is an $\mathscr{F}(P)$-module. If U such that E_U is isomorphic to a product, $\Gamma(E_U)$ is a free module.

<u>Definition</u>. A *linear connection* on the vector bundle E is as defined as an R-bilinear map

$$\mathscr{V}(P) \times \Gamma(E) \to \Gamma(E)$$

$$(V, \gamma) \to \nabla_V \gamma ,$$

satisfying the following conditions:

$$\nabla_{(fV)} \gamma = f \nabla_V \gamma \tag{2.1}$$

$$\nabla_V (f\gamma) = V(f)\gamma + f \nabla_V \gamma \tag{2.2}$$

for $V \in \mathscr{V}(P)$, $\gamma \in \Gamma(E)$.

Let us work locally, and suppose that $\Gamma(E)$ is a free $\mathscr{F}(P)$-module, i.e., that it has a basis

$$(\gamma_a)$$

$$1 \leq a, b \leq n = \text{dimensional fiber of } E$$

such that each $\gamma \in \Gamma(E)$ can be written in the form

$$\gamma = f^a \gamma_a \qquad (2.3)$$

with $f^1, \ldots, f^m \in \mathcal{F}(P)$.

(Summation convention is in force.)

Let (p^i) be a coordinate system for P. The vector fields ∂_i, defined such that

$$\partial_i(p^j) = \delta_i^j$$

are a basis for the $\mathcal{F}(P)$-module $\mathcal{V}(P)$. A linear connection then determines functions c_a^b such that

$$\nabla_{\partial_i}(\gamma_a) = c_{ai}^b \gamma_b . \qquad (2.4)$$

Then, (c_{ai}^b) are the *components* of the connection in terms of this "moving frame" and determine the connections.

Let V be an arbitrary vector field on P. Suppose that

$$V = v^i \partial_i ,$$

$$v^i \in \mathcal{F}(P) ,$$

i.e., the (v^i) are the components of V in these coordinates. Then, using (2.1):

$$\nabla_V(\gamma) = v^i \nabla_{\partial_i}(\gamma) . \qquad (2.5)$$

In particular, notice that if v is an arbitrary element of T(P), $v \in P_p$, then we can define

$$\nabla_v \gamma$$

as an element of the fiber of E above the point p. Choose *any* vector field V such that $V(p) = v$, and set

$$\nabla_v \gamma = (\nabla_V \gamma)(p) . \qquad (2.6)$$

This is the *covariant derivative of the cross-section* γ *at the tangent vector* v.

Suppose $t \to p(t)$ is a curve in P. Let $t \to \dot{p}(t)$ denote its tangent vector field. Let γ be an element of $\Gamma(E)$. Then,

$$t \to \gamma(\underline{p}(t))$$

and

$$t \to \nabla_{\underline{\dot{p}}(t)} \gamma(\underline{p}(t))$$

are mappings of the real numbers into vector bundles. The second one is said to be the *covariant derivative of the first along the curve* p. This will be the basic operation in our generalization of system theory. Let us sum up as follows.

Theorem 2.1. Let E be a vector bundle over a manifold P. For each curve σ in P consider the vector bundle E_σ over the real numbers, which assigns to each real t the vector space $\pi^{-1}(\sigma(t))$. (E_σ is the "pull-back" of the bundle E into the mapping σ: R → P.) Then, there is a mapping $\Gamma(E_\sigma) \to \Gamma(E_\sigma)$ induced by the connection, called *covariant derivative along the parameterized curve* σ.

3. GENERALIZED LINEAR INPUT-OUTPUT SYSTEMS AND COVARIANT DERIVATIVES

We are now prepared to formulate the basic system-theoretic ideas. Let P be a manifold and let E be a vector bundle over P with a fixed connection. For each p ∈ P, the fiber $\pi^{-1}(p)$ is a vector space, which we denote as X(p) and which we think of as "the state space over the point p." Let U and Y be vector spaces. Define vector bundles $\mathcal{A}, \mathcal{B}, \mathcal{C}$ over P defined as follows:

For p ∈ P, the fiber

$\mathcal{A}(p)$ is the space of linear maps A: X(p) → X(p)

$\mathcal{B}(p)$ is the space of linear maps B: U → $\mathcal{A}(p)$

$\mathcal{C}(p)$ is the space of linear maps: X(p) → Y .

U and Y are the familiar input and output spaces, which are fixed, but the "state space" X(p) depends on the points of P. To define a "linear system" in this set-up means to be given the following data:

a) A cross-section $\underline{A}$: p → A(p) of the bundle $\mathcal{A}$,
b) A cross-section $\underline{B}$: p → B(p) of the bundle $\mathcal{B}$,
c) A cross-section $\underline{C}$: p → C(p) of the bundle $\mathcal{C}$.

Given a curve

t → p(t)

on the base space P, we can now define the input-output relations as follows:

$$\nabla_{\dot{p}(t)} x(t) = A(p(t))(x(t)) + B(p(t))(u(t)) \tag{3.1}$$

$$y(t) = C(p(t))x(t) \quad .$$

This is to be solved for a cross-section

$$t \to x(t) \in X(p(t))$$

of the bundle E pulled back to the real numbers by the map. It is a first order, linear differential equation in the usual sense, but now with time-varying coefficients.

Thus, with zero initial conditions, (3.1) defines a set of linear input-output relations:

$$(\text{curves in } U) \to (\text{curves in } Y)$$

in the usual way. However, this input-output map depends on the curve $\underline{p}$. This is what enables us to set up a coupling between input-output relations and geometric structures in P.

4. LINEAR SYSTEMS COUPLED TO GEOMETRIC STRUCTURES

Keep the notation of Section 3. In addition to the linear system equations (3.1), consider a set of equations for the curve $t \to p(t)$ of the following form:

$$\dot{p}(t) = F(p(t)), x(t), u(t)) \quad . \tag{4.1}$$

In order to understand what F is, consider the product space $E \times U$ as a bundle over P: (x,u), $x \in X(p)$, $u \in U$, is mapped to p. F is then a fiber-preserving mapping of the bundle E' to the tangent bundle $T(p)$ to P.

Equations (3.1) and (4.1) *together* constitute the generalization we have in mind that is implicit in Kron's work, a coupling between "system theory" and "geometry".

References

1. G. Kron, Non-Riemannian geometry of rotating electrical machinery, *J. Math. Mech.* <u>13</u>, 103-194 (1934).

2. B. Hoffman, Kron's non-Riemannian electrodynamics, *Rev. Modern Phys.* <u>21</u>, 535-540 (1949).

3. K. Kondo, *The RAAG Memoirs*, Tokyo, 1955.

4. R. Hermann, *Differential Geometry and the Calculus of Variations*, Academic Press, 1968.

5. R. Hermann, *Vector Bundles in Mathematical Physics*, W.A. Benjamin, 1970.

Chapter 16

QUASI-COORDINATES AND MOVING FRAMES FOR LAGRANGIAN MECHANICAL SYSTEMS

1. INTRODUCTION

   In the work of Kron and the RAAG Memoirists, a valient effort was made to go beyond the 19th century methods of analytical mechanics and develop a geometric and "covariant" formalism according to the pattern of *Tensor Analysis*. The physical and engineering motivation for this--which remains strong today--is that the study of "complicated" mechanical systems (e.g., rotating electrical machinery) intrinsically involves complicated manifolds, which have no canonically given coordinate systems.

   A half-way house in this classical theory is what are called "quasi-coordinates". As I have pointed out in previous work (i.e., Volume 20), there is a direct link to Cartan's "method of the moving frame". What is involved is expressing the Pfaffian system on the one-jet bundle $J(R,Q)$ over the configuration manifold Q in terms of a basis of one-forms canonically defined by a basis of one-forms (a "moving frame") for Q.

   In this chapter, I will go over this basic material again, describing the equations of motion in terms of bases of differential forms for the configuration space.

2. LAGRANGIAN MECHANICAL SYSTEMS ON MANIFOLDS

   Let Q be a manifold, called *configuration space*. T denotes an interval of real numbers parameterized by t. $T(Q)$ denotes the tangent bundle to Q. Let

   $$J^1(T,Q)$$

   be the space of one-jets of maps from $T \to Q$. We have:

   $$J^1(T,Q) \equiv T \times T(Q) \tag{2.1}$$

   Given a curve

   $$t \to q(t) \quad ,$$

   it determines a *prolongation curve*

   $$t \to \left(q(t), \frac{dq}{dt}, t\right) \tag{2.2}$$

   in $T(Q) \times T \equiv J^1(T,Q)$. Thus,

$$t \to \left(q(t), \frac{dq}{dt}\right) \in T(Q)$$

is the *tangent vector curve* to the original curve in Q, while the curve (2.2) is the *graph* of the tangent vector curve.

The basic space for mechanics is

$$J^1(T,Q) \ .$$

A *mechanical system* is determined by three geometric objects:

a) A real-valued function

$$L: J^1(T,Q) \to R$$

called the *Lagrangian*.

b) A *Pfaffian system* $\mathcal{P}$ on $J^1(T,Q)$, i.e., a collection of one-forms which form a module over the ring of functions.

c) A one-differential form $\underline{F}$ on $T(Q) \times T$, called the *force-form*.

Suppose $\mathcal{P}$ and L. are given. L determines a one-form, denoted by θ, called the *Cartan form*. A tangent vector v to

$$J^1(T,Q) \equiv T \times T(Q)$$

is said to be a *d'Alembert vector* if

$$v \lrcorner d\theta - \underline{F} \in \mathcal{P} \ . \qquad (2.3)$$

(In words, (2.3) means that the one-covector $v \lrcorner d\theta - \underline{F}$ to $T(Q) \times T$ lies in the values of the differential forms in P at the point to which v is attached.)

<u>Definition</u>. A curve $t \to q(t)$ in Q is said to be an *orbit* of the mechanical system if its prolonged curve (2.2) satisfies the following conditions:

a) The tangent vectors to the prolonged curve are d'Alembert vectors

b) The prolonged curve (2.2) is an orbit curve of $\mathcal{P}$, i.e., its tangent vector curve $t \to v(t)$ satisfies the following condition:

$$v(t) \lrcorner \mathcal{P} = 0 \ . \qquad (2.4)$$

Quasi-Coordinates

3. QUASI-COORDINATES AND BASES OF DIFFERENTIAL FORMS/MOVING FRAMES

A set

$$(V_i^t), \quad i \leq i, j \leq n = \dim Q$$

of time-dependent vector fields on Q, which are *linearly independent*, define a system of *quasi-coordinates for* Q.

<u>Remark</u>. This is the classical terminology. I am not too happy with it, and will gradually phase it out of the notation. Note, however, that it is still used in physics and engineering books. A more appropriate term would be (following Cartan) to call it a *moving frame*.

Such vector fields determine convenient coordinates for $T(Q) \times T$. Let

$$\omega_t^i$$

be the dual differential one-forms on Q, which are dual to the V_i^t, i.e.,

$$\omega_t^i(V_j^t) = \delta_j^i \tag{3.1}$$

for each t.

For fixed $t \in T$, (ω_t^i) forms a *basis for one-forms on* Q.

<u>Notational Convention</u>: Consider the projection

$$\pi: J^1(T,Q) \to Q.$$

Pull the forms ω_t^i back via π^*. Denote them by the notation ω^i.

The ω_t^i also define *functions* on $J^1(T,Q)$, which are denoted by

$$y^i.$$

Namely,

$$y^i(v,t) = \omega_t^i(v) \tag{3.2}$$

for $v \in T(Q)$, $t \in T$.

Now, the one-forms

$$\omega^i, \, dy^i, \, dt \tag{3.3}$$

form a basis for one-forms on $J^1(T,Q)$.

<u>Remark</u>. It is usually the functions y^i defined by (3.2) that are called the "quasi-coordinates".

We can now use the bases (3.3) for one-forms to write down the equations (2.3) for the d'Alembert vectors. Set

$$\eta^i = \omega^i - y^i \, dt \tag{3.4}$$

The η^i are called the *contact forms*.

The Cartan form θ is now defined by the following formula:

$$\theta = L \, dt + \frac{\partial L}{\partial y^i} \eta^i \tag{3.5}$$

Let us now work out $d\theta$.

$$dL = \frac{\partial L}{\partial \omega^i} \omega^i + \frac{\partial L}{\partial y^i} dy^i + \frac{\partial L}{\partial t} dt \tag{3.6}$$

Now, the ω^i are no longer the differentials of functions. (This is, of course, why they are called "quasi-coordinates".) Hence, there are relations of the following form:

$$d\omega^i = c^i_{jk} \, \omega^j \wedge \omega^k + c^i_j \, \omega^j \wedge dt \tag{3.7}$$

The (c^i_{jk}, c^i_j) are functions on $Q \times T$, called the *structure functions* of the system of quasi-coordinates (or, moving frames).

Calculate using (3.4) and (3.7):

$$\begin{aligned}
d\theta^i &= d\omega^i - dy^i \wedge dt \\
&= c^i_{jk} \, \omega^j \wedge \omega^k + c^i_j \, \omega^j \wedge dt - dy^i \wedge dt \\
&= c^i_{jk} (\eta^j + y^j dt) \wedge (\eta^k + y^k dt) + c^i_j \, \eta^j \wedge dt - dy^i \wedge dt \\
&= c^i_{jk} \, \eta^j \wedge \eta^k + 2 c^i_{jk} \, y^k \eta^j \wedge dt - dy^i \wedge dt + c^i_j \, \eta^j \wedge dt
\end{aligned} \tag{3.8}$$

Quasi-Coordinates

Now we can calculate $d\theta$:

$$d\theta = dL \wedge dt + d\left(\frac{\partial L}{\partial y^i}\right) \wedge \eta^i + \frac{\partial L}{\partial y^i} d\eta^i$$

$$= \frac{\partial L}{\partial \omega^i} \eta^i \wedge dt + \frac{\partial L}{\partial y^i} dy^i \wedge dt + d\left(\frac{\partial L}{\partial y^i}\right) \wedge \eta^i$$

$$+ \frac{\partial L}{\partial y^i}(c^i_{jk}\eta^j \wedge \eta^k + 2c^i_{jk}y^k\eta^j \wedge dt - dy^i \wedge dt + c^i_j \eta^j \wedge dt)$$

$$= \frac{\partial L}{\partial \omega^i} \eta^i \wedge dt + d\left(\frac{\partial L}{\partial y^i}\right) \wedge \eta^i$$

$$+ \frac{\partial L}{\partial y^i}(c^i_{jk}\eta^j \wedge \eta^k + 2c^i_{jk}y^k\eta^j \wedge dt - c^i_j \eta^j \wedge dt)$$

Remark. The key fact in this calculation is that the terms involving dy^i have cancelled.

Set:

$$\alpha_i = -\frac{\partial L}{\partial \omega^i} dt + d\left(\frac{\partial L}{\partial y^i}\right) - 2\frac{\partial L}{\partial y^j} c^j_{ik} y^k dt + \frac{\partial L}{\partial y^j} c^j_i dt \qquad (3.10)$$

We now have the following basic formula:

$$\boxed{d\theta = \alpha_i \wedge \eta^i + \frac{\partial L}{\partial y^i} c^i_{jk} \eta^j \wedge \eta^k} \qquad (3.11)$$

Remark. In the case the "quasi-coordinates" are actually "coordinates", the third and fourth terms on the right hand side of (3.10) and the second term on the right hand side of (3.11) *vanish*. These terms are the generalizations of the "Coriolis forces" in elementary mechanics.

We can now present the main result of this chapter.

Theorem 16.1. Let $t \to q(t)$ be a curve in Q and let

$$t \to \sigma(t)$$

be its prolongation to be a curve in $J^1(T,Q)$. Then, $t \to q(t)$ is an orbit of the mechanical system

$$(Q, \theta, \underline{F}, \mathscr{P})$$

if and only if

$$\alpha_i(\sigma(t))\theta^i - \underline{F}(\sigma(t)) \in \mathscr{P}(\sigma(t)) \qquad (3.12)$$

for all t

where $t \to \sigma'(t)$ denotes the tangent vector curve to σ.

The <u>proof</u> follows from (3.11). Since $t \to \sigma(t)$ is the prolongation of a curve of Q, it is a solution curve of the Pfaffian system generated by the (θ^i), i.e.,

$$\theta^i(\sigma'(t)) = 0 . \qquad (3.13)$$

Now, the condition (2.1) means that

$$\sigma'(t) \lrcorner \, d\theta - \underline{F}(\sigma(t)) \in \underline{P}(\sigma(t)) \qquad (3.14)$$

Applying $\sigma'(t) \lrcorner$ to both sides of (3.11), and taking into account (3.13), proves (3.12).

Chapter 17

CYCLIC COORDINATES AND LINEAR SYSTEMS

1. INTRODUCTION

The theory of "cyclic" coordinates was a major topic in 19th century analytical mechanics (for example, in the work of Helmholtz, Routh and Whittaker). It is very important for modern work from the structural point of view--from certain systems governed by Lagrange equations, other systems can be obtained as "quotients". For example, geometric structure of electric circuit theory and certain aspects of the systems called "Toda lattices", is dominated by such considerations.

Helmholtz developed an approach to thermodynamics based on the cyclic coordinate formalism. (This is almost completely unknown today. I heard of it through Felix Klein's *Development of Mathematics in the 19th Century*. Math Sci Press is preparing a translation and revival of Helmholtz's work.)

In this paper I will survey certain aspects of this phenomenon--both in the most "modern" manifold-geometric framework and in the more classical language still used by engineers and physicists.

Another feature of this paper is the introduction of a general geometric form of "Kirkhoff's laws" of electrical network theory as a form of *nonholomonic constraints*.

2. SYMMETRIES OF MECHANICAL SYSTEMS

A *mechanical system* can be considered as a triple consisting of a manifold X, a two-differential form ω (not necessarily closed), and a vector field V such that

$$V \lrcorner \omega = 0 . \tag{2.1}$$

A *symmetry* of such a system is a diffeomorphism

$$\phi: X \to X$$

such that:

$$\phi_*(V) = V \tag{2.2}$$

$$\phi^*(\omega) = \omega \;. \tag{2.3}$$

An *infinitesimal symmetry* is a vector field W such that

$$\mathscr{L}_W(\omega) = 0 \tag{2.4}$$

$$[W,V] = 0 \;. \tag{2.5}$$

3. QUOTIENT MAPS FOR MECHANICAL SYSTEMS WITH GROUPS OF SYMMETRIES

Consider two mechanical systems

$$(X, \omega, V)$$

$$(X', \omega', V'') \;.$$

Suppose that G is a transformation group on the manifold X and that each $g \in G$ acts as a symmetry of the mechanical system. Let

$$\pi: X \to X'$$

be a map between the manifolds X and X'. We shall say that π is a quotient map (relative to the mechanical systems and the group G of symmetries) if the following conditions are satisfied:

a) $\pi_*(V) = V'$, i.e., π maps a time-trajectory of the mechanical system in X into a time-trajectory of the mechanical system on X'.

b) The orbits of G are the fibers of π.

c) $\pi^*(\omega') - \omega$ lies in the Grassmann ideal of differential forms on X which are zero on the fibers of π.

4. CYCLIC COORDINATES IN THE CLASSICAL SENSE FOR HAMILTONIAN SYSTEMS

In many of the examples in the classical literature the groups are abelian, and the decomposition maps π represent "ignorable coordinates". Let us review this situation in the simplest case, two degrees of freedom,

$$q = (q_1, q_2) .$$

Let $p = (p_1, p_2)$ be the dual momentum variables. Let

$$(p,q) \to h(p,q)$$

be the *Hamiltonian* function. *Hamilton's equations* are:

$$\frac{dq}{dt} = h_p$$

$$\frac{dp}{dt} = -h_q . \tag{4.1}$$

The coordinates (q_1) is said to be *cyclic* if h does not depend on q_1, i.e., if

$$h_{q_1} = 0 . \tag{4.2}$$

Equations (4.1) now imply that:

$$\frac{dp_1}{dt} = 0 ,$$

i.e., the function p_1 is constant on the trajectories.

Let X be the space of the variables (p,q,t), i.e., X is (locally) R^5. Set

$$\omega = dp \wedge dq - dh \wedge dt \tag{4.3}$$

$$V = h_p \frac{\partial}{\partial q} - h_q \frac{\partial}{\partial p} + \frac{\partial}{\partial t} \ . \tag{4.4}$$

Then,

$$V \lrcorner \ \omega = 0 \ , \tag{4.5}$$

i.e., V is *Cauchy characteristic* for the exterior differential system generated by ω.

Let X' be the space of variables (q_2, p_1, p_2, t). Let

$$\pi: X \to X'$$

be the following projection map:

$$\pi(q_1, q_2, p_1, p_2, t) = (q_2, p_1, p_2, t) \ . \tag{4.6}$$

Set:

$$\omega' = dp_2 \wedge dq_2 - dh \wedge dt \ . \tag{4.7}$$

ω' is then a closed two-form on X', *since h does not depend on* q_1. Set:

$$V' = h_{p_2} \frac{\partial}{\partial q_2} - h_{q_2} \frac{\partial}{\partial p_2} + \frac{\partial}{\partial t} \ . \tag{4.8}$$

Then,

$$V' \lrcorner \ \omega' = 0 \ . \tag{4.9}$$

Because of the cyclicity condition (4.2), V reduces to the following form:

$$V = h_{p_1} \frac{\partial}{\partial q_1} + h_{p_2} \frac{\partial}{\partial q_2} - h_{q_2} \frac{\partial}{\partial p_2} + \frac{\partial}{\partial t} \ . \tag{4.10}$$

Now,

Cyclic Coordinates

$$\pi_*\left(\frac{\partial}{\partial q_1}\right)(f) = \frac{\partial}{\partial q_1}(\pi^*(f)) . \qquad (4.11)$$

But $\pi^*(f)$ depends only on (q_2, p_1, p_2, t), i.e., the right hand side of (4.10) is zero. Hence,

$$\pi^*\left(\frac{\partial}{\partial q_1}\right) = 0 . \qquad (4.12)$$

Therefore,

$$\pi_*(V) = V' . \qquad (4.13)$$

We have then proved:

Theorem 4.1. With the above definitions, π defines a quotient map between the mechanical system (X, ω, V) and (X', ω', V'). The group G which acts on the fibers of π as a group of symmetries of the mechanical system (X, ω, V) is that generated by the vector field

$$\frac{\partial}{\partial q_1} .$$

In fact, notice that the very property of "cyclicity" with respect to q_1 is the "symmetry" condition

$$\left[\frac{\partial}{\partial q_1}, V\right] = 0 .$$

5. COMPLETELY DEGENERATE LAGRANGIANS

Let Q be an n-dimensional manifold with coordinates (q^i), $(q, v) \equiv (q', v')$ the corresponding coordinates of the tangent bundle $T(Q)$ and

$$(q, v, t) \equiv (q^i, v^i, t)$$

the corresponding coordinates of the one-jets

$$J^1(T, Q) .$$

(The summation convention is in force. We revert to the tensor analysis notation in this section because it is computationally convenient.)

A *Lagrangian* is a real valued function $L(q,v,t)$ of these coordinates. Let

$$d\theta(L) = L dt + L_{v^i}(dq^i - v^i dt)$$

be the Cartan form. It was proved in *Geometry, Physics and Systems* that every curve $t \to q(t)$ is a solution of Lagrange's equations if and only if

$$d\theta(L) = 0 \ . \tag{5.1}$$

If this is so, we shall say that the Lagrangian is *completely degenerate*. (In other words, Lagrange's equations are *identities*.) Let us work out the conditions on L that (5.1) be satisfied:

$$d\theta(L) = \left(d\left(L_{v^i}\right) - L_{q^i} dt \right) \wedge (dq^i - v^i dt) \ . \tag{5.2}$$

By the Cartan lemma, $d\theta(L)$ is zero if and only if there are functions γ_{ij} such that:

$$dL_{v^i} - L_{q^i} dt = \gamma_{ij}(dq^j - v^i dt) \ , \tag{5.3}$$

$$\gamma_{ij} = \gamma_{ji} \ . \tag{5.4}$$

(5.3) requires first that:

$$L_{v^i v^j} = 0 \ , \tag{5.5}$$

i.e., L is a linear function of v. L_{v^i} is a function of (q,t) above.

$$\left(L_{v^i}\right)_{q^j} = \gamma_{ij}$$

Condition (5.4) now requires that there is a function

$$\alpha(q,t)$$

such that

$$L_{v^i} = \alpha_{q^i} .$$

Hence,

$$\left(L - \alpha_{q^j} v^j\right)_{v^i} = 0 ,$$

i.e.,

$$L = \alpha_{q^j} v^j + \beta(q,t) \qquad (5.6)$$

$$\gamma_{ij} = \alpha_{q^i q^j} \qquad (5.7)$$

Finally,

$$\alpha_{q^i t} - \alpha_{q^j q^i} v^j - \beta_{q^i} = -\alpha_{q^i q^j} v^j ,$$

or

$$\alpha_{q^i t} = \beta_{q^i} ,$$

or

$$(\alpha_t - \beta)_{q^i} = 0 ,$$

$$\alpha_t = \beta + \delta(t) .$$

Thus, we have proved:

<u>Theorem 5.1</u>. The Lagrangian equation associated to the Lagrangian L are identities if and only if L is of the following form:

$$L = \alpha(q,t)_{q^i} v^i + \alpha_t + \delta(t) , \qquad (5.8)$$

where α and δ are arbitrary functions of the indicated variables.

Proof. We have proved one direction. Let us check directly that for L of form (5.8), Lagrange's equations are identities:

$$\frac{d}{dt}\left(L_{v^i}\right) - L_{q^i} = \frac{d}{dt}\left(\alpha_{q^i}\right) - \alpha_{q^i q^j}\frac{dq^j}{dt} - \alpha_{tq^i} \quad ,$$

which is indeed identically zero for every curve $t \to q(t)$.

6. CYCLICITY FROM THE POINT OF VIEW OF LAGRANGE'S EQUATIONS

The "Hamiltonian" approach to cyclicity is the one most studied in the contemporary literature. However, the more traditional Lagrangian approach is, in my opinion, more fundamental. Let us examine the "cyclicity" question in this framework. Again for simplicity, we shall restrict attention to systems with two degrees of freedom.

Let

$$q = (q_1, q_2)$$

be the configuration coordinates, $v = (v_1, v_2)$ the tangent vector coordinates. The Lagrangian $L(q,v)$ is a function of these coordinates.

By the logic of the situation we have:

Theorem 6.1. q_1 is a cyclic coordinate for the Lagrangian L, i.e., q_1 does not occur explicitly in L if and only if L_{q_1} is a degenerate Lagrangian in the sense of Section 5.

Let us work out the condition that q_1 be cyclic for L more explicitly using Theorem 5.1. L_{q_1} is of the form

$$L_{q_1} = \alpha_{q_1} v_1 + \alpha_{q_2} v_2 + \alpha_t + \delta \quad , \tag{6.1}$$

where α is a function of (q,t) and δ a function of t.

Theorem 6.2 (Routh). q_1 is a cyclic coordinate for L if and only if L is of the following form:

Cyclic Coordinates

$$L = \alpha_{q_1} v_1 + \alpha_{q_2} v_2 + \alpha_t + \delta(t) q_1 + K(q_2, v_1, v_2) , \qquad (6.2)$$

where α is an arbitrary function of (q_1, q_2, t), δ an arbitrary function of t, and K an arbitrary function of (q_2, v_1, v_2).

Proof. Again, let us check this by explicitly writing down Lagrange's operators and checking that q_1 does not appear explicitly:

$$\frac{d}{dt}\left(L_{v_1}\right) - L_{q_1} = \frac{d}{dt}\left(\alpha_{q_1} + K_{v_1}\right) - \alpha_{q_1 q_1}\frac{dq_1}{dt} - \alpha_{q_1 t} - \delta(t)$$

$$= \frac{d}{dt}\left(K_{v_1}\right) - \delta(t) \qquad (6.3)$$

$$\frac{d}{dt}\left(L_{v_2}\right) - L_{q_2} = \frac{d}{dt}\left(\alpha_{q_2} + K_{v_2}\right) - \alpha_{q_1 q_2}\frac{dq_1}{dt} - \alpha_{q_2 q_2}\frac{dq_2}{dt}$$

$$- K_{q_2} - \alpha_{t q_2}$$

$$= \frac{d}{dt}\left(K_{v_2}\right) - K_{q_2} \qquad (6.4)$$

Thus, we see the beautiful property of the Routh theory—the Lagrange equations in two degrees of freedom is reduced to another of one degree of freedom, namely (6.4), plus a constraint determined by (6.3).

7. SYSTEMS WITH TWO DEGREES OF FREEDOM WHOSE CONFIGURATION COORDINATES ARE CYCLIC

So far, we have determined Lagrangians which have one cyclic coordinate. At the other extreme are those for which all coordinates are cyclic. Let us consider this situation in the simplest case of two degrees of freedom

$$q = (q_1, q_2) .$$

Suppose $L(q,v)$ is a Lagrangian such that both q_1 and q_2 are cyclic.

By Theorem 6.2 L is of the following form:

$$L = \alpha_{q_1} v_1 + \alpha_{q_2} v_2 + \alpha_t + \delta(t)q_1 + K(q_2, v_1, v_2)$$

$$= \beta_{q_1} v_1 + \beta_{q_2} v_2 + \beta_t + \gamma(t)q_2 + J(q_1, v_1, v_2) \quad . \tag{7.1}$$

Differentiate both sides of (7.1) with respect to q_1:

$$\alpha_{q_1 q_1} v_1 + \alpha_{q_2 q_1} v_2 + \alpha_{q_1 t} + \delta = \beta_{q_1 q_1} v_1 + \beta_{q_2 q_1} v_2 + \beta_{t q_1} + J_{q_1}$$

We see that J_{q_1} is linear in v_1 and v_2. Suppose for simplicity that L is free of explicit time dependence. Thus, J is of the form

$$J = A_1 v_1 + A_2 v_2 + A_3(v_1, v_2) \quad , \tag{7.2}$$

where A_1 and A_2 are functions of (q_1, q_2).

Theorem 7.1. In order that the time-independent Lagrangian L be cyclic with respect to both q_1 and q_2, it must be of the form:

$$L = Av_1 + Bv_2 + C(v_1, v_2) \quad , \tag{7.3}$$

where A and B are functions of (q_1, q_2).

We must now determine the additional conditions cyclicity puts on A and B. Let us do this directly, by writing down Lagrange's equations:

$$\frac{d}{dt}\left(L_{v_1}\right) - L_{q_1} = \frac{d}{dt}\left(A - C_{v_1}\right) - A_{q_1} v_1 - B_{q_1} v_2$$

$$= A_{q_1} v_1 + A_{q_2} v_2 - C_{v_1 v_1} \frac{dv_1}{dt} - C_{v_1 v_2} \frac{dv_2}{dt} - A_{q_1} v_1 - B_{q_1} v_2$$

$$= \left(A_{q_2} - B_{q_1}\right) v_2 - C_{v_1 v_1} \frac{dv_1}{dt} - C_{v_1 v_2} \frac{dv_2}{dt} \quad . \tag{7.4}$$

The cyclicity condition is that:

Cyclic Coordinates

$$A_{q_2} - B_{q_1} = k_1 \, , \tag{7.5}$$

a constant. Then, $t \to (v_1(t), v_2(t))$ satisfies the following equation:

$$k_1 v_2 - C_{v_1 v_2} \frac{dv_2}{dt} - C_{v_1 v_1} \frac{dv_1}{dt} = 0 \, . \tag{7.6}$$

Permute v_1 and v_2 to obtain the other equation

$$A_{q_1} - B_{q_2} = k_2 \tag{7.7}$$

$$k_2 v_2 - C_{v_2 v_1} \frac{dv_1}{dt} - C_{v_2 v_2} \frac{dv_2}{dt} = 0 \, . \tag{7.8}$$

Theorem 7.2. Lagrange's equations with two degrees of freedom for which the configuration coordinates are both cyclic are nonlinear equations of the following form for $v = (v_1, v_2)$:

$$\boxed{\begin{aligned} C_{v_1 v_1} \frac{dv_1}{dt} + C_{v_1 v_2} \frac{dv_2}{dt} &= k_1 v_1 \\ C_{v_2 v_2} \frac{dv_2}{dt} + C_{v_2 v_1} \frac{dv_1}{dt} &= k_2 v_2 \end{aligned}} \tag{7.9}$$

C is an arbitrary function of (v_1, v_2), k_1 and k_2 are constants.

8. LINEAR CYCLICITY

Let us combine the concepts of "cyclicity" and "linearity". Let Q be R^n. Denote n-vectors by columns:

$$q = \begin{pmatrix} q_1 \\ \vdots \\ q_n \end{pmatrix}$$

$$q' = (q_1, \ldots, q_n) \quad .$$

v denotes another n-vector. $\begin{pmatrix} q \\ v \end{pmatrix}$ is a partitioned 2n-vector. Suppose

$$L = \frac{1}{2} (q', v') A \begin{pmatrix} q \\ v \end{pmatrix} \qquad (8.1)$$

$$A = \begin{pmatrix} A_{11} & A_{12} \\ A_{21} & A_{22} \end{pmatrix} \qquad (8.2)$$

A is a symmetric $2n \times 2n$ matrix, partitioned as indicated into $n \times n$ submatrices. The symmetry conditions are:

$$A' = \begin{pmatrix} A'_{11} & A'_{21} \\ A'_{12} & A'_{22} \end{pmatrix}$$

$$= A \quad ,$$

or

$$A'_{11} = A_{11}$$

$$A'_{22} = A_{22} \qquad (8.3)$$

$$A'_{21} = A_{12} \quad .$$

Then,

Cyclic Coordinates

$$L_v = (0,1) \, A \begin{pmatrix} q \\ v \end{pmatrix}$$

$$= (0,1) \begin{pmatrix} A_{11}q + A_{12}v \\ A_{21}q + A_{22}v \end{pmatrix}$$

$$= A_{21}q + A_{22}v$$

$$L_q = (1,0) \, A \begin{pmatrix} q \\ v \end{pmatrix}$$

$$= A_{11}q + A_{12}v \ .$$

The Lagrange operator is:

$$v \to \frac{d}{dt}(L_v) - L_q \equiv \Delta_L(q) \ .$$

The cyclicity condition is then:

$$A_{11} = 0 \ . \tag{8.5}$$

Δ_L then takes the following form:

$$\Delta_L(v) = A_{21}v + A_{22}v_t - A_{12}v$$

$$= (A_{21} - A_{12})v + A_2 v$$

$$= \text{, in view of relations (8.3)}$$

$$(A'_{12} - A_{12})v + A_{22}v \ . \tag{8.6}$$

Let us summarize in the following form, with a slight change in notation:

<u>Theorem 8.1</u>. Let L be a quadratic time invariant Lagrangian for R^n, for which the configuration coordinate is cyclic. Then, the corresponding linear Lagrange differential operator on the velocity vector takes the following form:

$$\Delta_L(v) = \alpha v_t + \beta v \;, \tag{9.7}$$

where v_t denotes the time-derivative. α and β are $n \times n$ matrices, β symmetric, α skew-symmetric.

9. THE LAGRANGE-RAYLEIGH EQUATIONS WITH EXTERNAL FORCES

Let Q be a manifold with coordinates (and the summation convention)

$$(q^i) \;, \quad 1 \leq i, j \leq n = \dim Q \;.$$

We revert to the classical notation and denote the tangent coordinates by

$$(\dot{q}_1)$$

$(q, \dot{q}, t)$ then are the coordinates of the Ehresmann jet space

$$J^1(R, Q) \;.$$

Let

$$\pi: J^1(R, Q) \to Q$$

be the projective map on Q, i.e., $\pi(q, \dot{q}, t) = q$. Let

$$E = \pi^{-1}(T^d(Q))$$

be the pull-back under π of the *cotangent bundle* $T^d(Q)$ on Q. Regard E as a vector bundle over $J^1(R, Q)$. The fiber of E above the point $(q, \dot{q}, t)$ is the cotangent space Q_q^d to Q at q.

A more convenient way of thinking of this is to regard $J^1(R, Q)$ as the set of ordered triples

$$(t, q, \dot{q}) \;,$$

where $t \in R$, $q \in Q$, and $\dot{q}$ is an element of the tangent space to Q above q, Q_q. E is then the ordered quadruples

$$(t, q, \dot{q}, u) \;,$$

Cyclic Coordinates

where

$$t \in R, \quad q \in Q, \quad \dot{q} \in Q_q, \quad u \in Q_q^d.$$

Let (q^i) be a given coordinate system of functions on Q. They define real-valued functions $\dot{q}^i$ and u_i on E by the rule:

$$\dot{q}^i(t,q,\dot{q},u) = dq^i(\dot{q})$$

$$u_i(t,\dot{q},u) = \langle u, \partial/\partial q^i \rangle$$

Let L and R be two real-valued functions on E, the *Lagrangian* and *Rayleighan*. A vector field V on E is a *Lagrange-Rayleigh vector field* if it satisfies the following conditions:

$$V(t) = 1$$

$$(\partial q^i - \dot{q}^i dt)(V) = 0$$

$$V\left(\frac{\partial L}{\partial \dot{q}_i}\right) - \frac{\partial L}{\partial q^i} = \frac{\partial R}{\partial \dot{q}_i} + u_i$$

These formulas say that the integral curves of V are solutions of the Lagrange-Raleigh equations, with external forces u_i. In classical notation they would be written as

$$\frac{d}{dt}(L_{\dot{q}}) - L_q = R_{\dot{q}} + u.$$

Of course an alternate approach is to define a two-form ω as follows:

$$\omega = \left(d\left(L_{\dot{q}_i}\right) - L_{q_i} dt\right) \wedge (dq^i - \dot{q}_i dt) - \left(R_{\dot{q}_i} + u_i\right) dt \wedge dq^i.$$

V then satisfies the following Cartanian equation:

$$V \lrcorner \, \omega = 0.$$

10. DIRECT PRODUCT OF LAGRANGE-RALEIGH SYSTEMS WITH EXTERNAL FORCES

We can formalize a construction which is implicit in classical electric circuit theory. Let $Q_1, \ldots, Q_m$ be configuration space manifolds of mechanical systems. Let $E_1, \ldots, E_n$ be the fiber bundles over $Q_1, \ldots, Q_m$ constructed in the previous section. (Thus, the fiber of this bundle is the coordinates of time, velocity, and external force. Recall that E is the pull back of the cotangent bundle of Q under the projection map $J^1(R,Q) \to Q$.)

Now, set

$$Q = Q_1 \times \cdots \times Q_m ,$$

the Cartesian product of the manifolds $Q_1, \ldots, Q_m$. The fiber spaces $E_1, \ldots, E_n$ on which the Lagrangian and Rayleighans "lie" may be identified with geometric objects constructed out of the tangent and cotangent bundles of the configuration manifolds:

$$E_1 = R \times (T(Q_1) \oplus T^d(Q_1)) .$$

($\oplus$ denotes "direct sum of vector bundles".) Thus, a point of E_1 may be thought of as an ordered quadruple

$$(t, q, \dot{q}, u) ,$$

with

$$t \in R, \quad q \in Q, \quad \dot{q} \in Q_q, \quad u \in Q_q^d .$$

Thus, the E-space associated with Q is the ordered triples of the form:

$$e = (t, q_1, \ldots, q_m, \dot{q}_1, \ldots, \dot{q}_m, u_1, \ldots, u_m) .$$

The total Lagrangian and Raleighian can then be defined as follows:

Cyclic Coordinates

$$R(e) = R_1(t,q_1,\dot{q}_1) + \cdots + R_m(t,q_m,\dot{q}_m)$$

$$L(e) = L_1(t,q_1,\dot{q}_1) + \cdots + L_m(t,q_m,\dot{q}_m) \ .$$

With the total u identified with $(u_1,\ldots,u_m)$ we see that the Lagrange-Rayleigh equations with external forces for L splits up into the direct sum of the Lagrange-Rayleigh equations of the subsystems

$$\frac{d}{dt}\left((L_1)_{\dot{q}_1}\right) - L_{1q_1} = R_{1\dot{q}_1} + u_1$$

$$\vdots$$

The "engineering method" of putting in interactions between these systems is to introduce *constraints*, i.e., submanifolds of E. (The "physicists method" would be to add "interaction terms".) For example, in circuit theory, these constraint submanifolds are "Kirkhoff's Laws".

We will now describe a general version.

11. LAGRANGE-RAYLEIGH SYSTEMS WITH NON-HOLONOMIC CONSTRAINTS

Let us first recall (from *Geometry, Physics and Systems*, for example) how *non-holonomic constraints* may be introduced into the Lagrangian mechanics.

Let Q continue as a manifold with coordinates $q = (q^i)$. Let $(q,\dot{q},t)$ denote the corresponding coordinates of the one-jet space. Let E denote the space of ordered quadruples

$$(t,q,\dot{q},u) \ ,$$

with

$$t \in R, \quad q \in Q, \quad \dot{q} \in Q_q, \quad u \in Q_q^d \ .$$

Let (q^i) be a coordinate system for Q. As explained previously, there is a coordinate system labelled

$$(t,q^i,\dot{q}^i,u_i)$$

for E. Let L and R be real-valued functions on $J^1(R,Q)$. Construct the two-forms

$$\omega = \left(d\left(L_{\dot{q}^i} \right) - L_{q^i}\, dt - R_{\dot{q}^i}\, dt - u_i\, dt \right) \wedge (dq^i - \dot{q}^i dt) \ .$$

Then, the vector fields V on E such that

$$V(t) = 1$$

$$V \lrcorner \, \omega = 0 \qquad (11.1)$$

$$V \lrcorner \, (dq^i - \dot{q}^i dt) = 0$$

define the Lagrange-Rayleigh equations with external forces, in the sense that their integral curves satisfy these equations.

Let θ^a, $0 \leq a, b \leq 1$, be a set of one-forms on E that are to be introduced as "constraints". Replace (11.1) with the following:

$$V(t) = 1$$

$$V \lrcorner \, \omega = \lambda_a \theta^a \qquad (11.2)$$

$$V \lrcorner \, (dq^i - \dot{q}^i dt) = 0 \ .$$

The integral curves of such V's are the orbits of the *constrained* system.

These constraints are *non-holonomic* if the Pfaffian system (θ^a) is not completely integrable, i.e., if

$$d\theta^a \wedge \theta^1 \wedge \cdots \wedge \theta^m \neq 0 \ . \qquad (11.3)$$

If it is *holonomic*, the system (11.2) is equivalent to a parameterized family of systems of type (11.1). $\theta^a = 0$ then defines a foliation of E. Relations (11.2) imply that

$$\theta^a(V) = 0 \ ,$$

i.e., V is tangent to the leaves of the foliation. Restrict V and ω to the leaves of this foliation to obtain a system of type (11.1).

However, in this work we are mainly interested in the non-holomonic case. We will now specify a special form of the constraints that we call "Kirkhoffian".

12. LINEAR LAGRANGE-RALEIGH SYSTEMS WITH KIRKHOFFIAN CONSTRAINTS

Keep the notation of Section 1. Let E, Q, and (q^i) be as before. Specialize L and R as follows:

$$L = \frac{1}{2} L_{ij} \dot{q}^i \dot{q}^j + \frac{1}{2} K_{ij} q^i q^j \tag{12.1}$$

$$R = \frac{1}{2} R_{ij} \dot{q}^i \dot{q}^j . \tag{12.2}$$

Then,

$$L_{\dot{q}^i} = L_{ij} \dot{q}^j \tag{12.3}$$

$$L_{q^i} = K_{ij} q^j \tag{12.4}$$

$$\omega = (L_{ij} d\dot{q}^j - K_{ij} q^j - dt - R_{ij} q^j dt - u_i dt) \wedge (dq^i - \dot{q}^i dt) \tag{12.5}$$

Introduce *Kirkhoffian constraints* as Pfaffian equations of the form:

$$\theta_a = (\alpha_{ai} dq^i + \beta^i_{ai} v_i dt) = 0 \tag{12.6}$$

The contact forms are

$$\eta^i = dq^i - \dot{q}^i dt .$$

Look for vector fields V on the space of variables $(t, q, \dot{q}, u)$ such that:

$$V(t) = 1$$
$$\eta^i(V) = 0 \tag{12.7}$$
$$\theta_a(V) = 0$$

$$V \lrcorner \, \omega = \lambda^a \theta_a \qquad (12.8)$$

for some functions λ^a.

Now,

$$V \lrcorner \, \omega = \left(V\!\left(L_{\dot q^i}\right) - \left(R_{\dot q^i} + u_i\right) \right) \eta^i = \lambda^a (\alpha_{ai} dq^i + \beta^i_a v_i dt) \; ,$$

or

$$V\!\left(L_{\dot q^i}\right) - \left(R_{\dot q^i} + v_i\right) = \lambda^a (\alpha_{ai}) \left(V\!\left(L_{\dot q^i}\right) - \left(R_{\dot q^i} + v_i\right) \right) \dot q^i$$

$$= - \lambda^a \beta^i_a v_i$$

Using the values given for L and R,

$$L_{\dot q^i} = L_{ij} \dot q^j$$

$$R_{\dot q^i} = R_{ij} \dot q^j \; .$$

Restrict these relations to a curve

$$t \to \left(q(t), \; \dot q(t) = \frac{dq}{dt} = q_t, \; v(t), \; \lambda(t) \right) \; ,$$

which is an integral curve of V_i. The following relations result:

$$L_{ij} q^j_{tt} - R_{ij} q^j_t - v_i = \lambda^a \alpha_{ai} \qquad (12.9)$$

$$(L_{ij} q^j_{tt} - T_{ij} q^j_t - v_i) q^i_t = - \lambda^a \beta^i_a v_i \; . \qquad (12.10)$$

$$\alpha_{ai} q^i_t - \beta^i_a u_i = 0 \qquad (12.11)$$

Notice that (12.10) is a consequence of (12.9) and (12.11). Hence, (12.9) and (12.11) are the only conditions. The integral curves of V, i.e., the equations of motion of the physical system defined by L and R, constrained by the Kirkhoff relations (12.11) are then of the following form, when written in matrix notation:

Cyclic Coordinates

$$Lq_{tt} - Rq_t - u = \lambda\alpha \;. \tag{12.12}$$

$$\alpha q_t + \beta v = 0 \;. \tag{12.13}$$

Notice that (12.12) and (12.13) are $(n+m)$ equations for the $(2n+m)$ unknowns q, u, λ. Then, generically, they can be written in "state space" form

$$\frac{dx}{dt} = Ax + (\text{input}) \;.$$

Explicitly, carrying out this reduction is algebraically highly non-trivial.

Chapter 18

DIFFERENTIAL GEOMETRY OF ENGINEERING-MECHANICS SYSTEMS

1. INTRODUCTION

One of the interesting mathematical developments of the last twenty years has been the revival within the context of professional mathematics of the classical subject of *analytical mechanics*. This has been "geometrized" and some classical problems have been profitably examined in a new light. The treatises by Abraham and Marsden [1] and Arnold [2] are now the standard places to sample this material. Above all, the links between analytical mechanics and quantum mechanics have been made much more precise.

However, this work has only penetrated into the traditional problems of *particle mechanics*. Not much has been done with the much more complicated problems of "engineering mechanics". Also, the new insights of system-control theory have not been brought in full force to bear on this area. This has hindered engineering development of this knowledge, since it is precisely the "control" aspects of the theory which are of most practical use. Thus, a program of study of the application of the geometric techniques to the more complicated mechanical system encountered in engineering seems indicated.

One of the most remarkable areas of work in the classical engineering-mechanics literature is that of Gabriel Kron and the RAAG group led by K. Kondo. Inspired by classical tensor analysis, they sketched out vast areas which could be understood in geometric terms, including electric machine and circuit theory, aircraft dynamics, and many others. What is most interesting is that they had a very systems-oriented point of view, emphasizing connections and interaction of mechanical systems and the decomposition ("tearing") of "large" systems into small ones.

The geometric tools available to Kron and Kondo were very primitive. I believe that with modern manifold theory and algebraic geometry, much more could be done. Symbolic computer systems such as MACSYMA also show promise of doing the often tedious calculations more readily and making the material more accessible to the working engineer. In this report I want to sketch some of these possibilities.

2. THE LAGRANGE EQUATIONS

Let Q be a manifold with coordinates $q = (q_1,\ldots,q_n)$. We use a hybrid notation combining the vectorial notation now in use in the engineering literature and the manifold theoretic notation of the geometers.

The *tangent bundle* $T(Q)$ has coordinates

$$(q,v) .$$

(v stands for "velocity". They are the coordinates of tangent vectors.)

Let $L(q,v)$ be a real valued function on $T(Q)$. Let F be a map: $T(Q) \to R^n$. In the classical style, *Lagrange's equations* take the form:

$$\frac{d}{dt}(L_v) - L_q = F .$$

$$\frac{dq}{dt} = v .$$

(2.1)

(Subscripts denote partial derivatives.)

Equations (2.1) (the *Newton-Lagrange equations*) are a set of ordinary, second-order differential equations. They can be put into a more coordinate-free form. Define one-differential forms on the space of variables (q,v,t) as follows:

Engineering Systems

$$\omega = dL_v - L_q dt - F dt \qquad (2.2)$$

$$\theta = dq - v dt$$

Theorem 2.1. The solutions of the equations (2.1) are the curves in the manifold $T(Q) \times R$ on which the one-forms (2.1) restrict to *zero*.

Another way of putting this is to look for the vector fields V on $T(Q) \times R$ such that:

$$V \lrcorner (\omega \wedge \theta) = F$$
$$V \lrcorner \theta = 0 \, . \qquad (2.3)$$

The solutions of (2.1) are then the *integral curves* (or *orbit curves*) of the vector field V.

This is an optimally "geometric" form of the laws of mechanics-- completely coordinate free. It is even more systematized when $T(Q) \times R$ identified with $J^1(R,Q)$, the *1-jets* of mappings of $R \to Q$.

The beauty of this formulation is that it might well be convenient to choose other coordinates for $T(Q) \times R$. Using the laws of the vector field-differential form calculus, this is easily done. (The engineering and physics literature is very confusing about this.)

For example, the *momenta* are just the functions

$$p = L_v \, .$$

Certain coordinates may be "cyclic" or "ignorable". This means that there is a map

$$\pi : T(Q) \to M$$

to another manifold M, and a two-form α on M such that:

$$\pi^*(\alpha) = \omega \wedge \theta$$

However, the full power of the formalism comes when *constraints* are considered.

3. LAGRANGIAN MECHANICS WITH CONSTRAINTS

Suppose a mechanical system

$$(Q, L, F)$$

is defined. *Constraints* may be defined by a set

$$\eta = (\eta_1, \ldots, \eta_m)$$

of one-forms on $T(Q) \times R$. Suppose

$$\eta = \alpha dq$$

$$\alpha = (\alpha_1, \ldots, \alpha_m)$$

The Newton-Lagrange equations *with constraints*, are then the following:

$$\frac{d}{dt}(L_v) - L_q = F + \lambda \alpha \quad,$$

$$\lambda = (\lambda_1, \ldots, \lambda_m)$$

$$\alpha(q) \frac{dq}{dt} = 0 \quad.$$

These equations are to be solved for a curve $t \to (q(t), \lambda(t))$ in $Q \times R^m$.

Here is another formulation. Let M be the space of variables

$$(q, v, t, \lambda)$$

Introduce the following two-form:

$$\Omega = (dL_v - L_q dt - Fdt - \lambda \alpha dt) \wedge (dq - vdt)$$

Consider vector fields V on this space such that:

$$V \lrcorner \Omega = 0$$

$$V \lrcorner \eta = 0 \quad.$$

The integral curves of these vector fields are the solutions of the Newton-Lagrange equations with constraints.

Engineering Systems

The constraints are *holomonic* if the Pfaffian system

$$\eta = 0$$

is *completely integrable* (in the Frobenius sense), i.e., if

$$d\eta \wedge \eta = 0 .$$

4. INPUT-OUTPUT THEORY FOR MECHANICAL SYSTEMS

The material in the last Section is essentially a souped-up version of 19th century mechanics. Modern system theory has added a new ingredient to this classical stew--the addition of *inputs* and *outputs*.

Now, an input-output (deterministic) system is usually taken to be one of the form:

$$\frac{dx}{dt} = f(x,u)$$
$$y = g(x) .$$
(4.1)

For the purposes of Lagrangian mechanics, it seems desirable to generalize this to systems defined implicitly:

$$f\left(\frac{dx}{dt}, x, u\right) = 0$$
$$y = g(x)$$
(4.2)

A Lagrangian mechanical system can be put into this form in many ways. The method that seems most reasonable from the physical-engineering point of view can be described as follows. Consider a Lagrangian system of the following form:

$$\frac{d}{dt}(L_v) - L_q = F .$$
(4.3)

Suppose the forms F are functions $F(q,v,u)$ of position, velocity, and input. Set

$$x = (q,v) \ .$$

Then equations (4.3) can be written in form (4.1), with appropriate choice of f.

These equations can be written in Pfaffian form in terms of the variables (x,u,λ,t). Perhaps this is the ultimate "geometric" form of the equations of system theory.

5. INTERACTION OF LAGRANGIAN SYSTEMS

One of the major reasons that physicists, to this day, use Lagrangian mechanics is that it provides a very natural way to introduce interactions of mechanical systems.

Consider two systems

$$(q,v,L,F)$$

$$(q',v',L',F') \ .$$

The uncoupled equations are:

$$\frac{d}{dt}(L_v) - L_q - F = 0 \tag{5.1}$$

$$\frac{d}{dt}(L'_{v'}) - L'_{q'} - F' = 0 \ .$$

Now, construct a Lagrangian for the composite system of the following form:

$$L''(q,v,q',v') = L(q,v) + L'(q',v') + K(q,q') \tag{5.2}$$

$$F'' = F + F' \ .$$

K is an "interaction" term. (It may, of course, be more complicated than this.)

We can construct the Lagrange equations associated with the Lagrangian L":

Engineering Systems

$$L''_v = L_v$$

$$L''_{v'} = L'_{v'}$$

$$L''_q = L_q + K_q$$

$$L''_{q'} = L'_{q'} + K_{q'}$$

$$\frac{d}{dt}(L''_v) - L''_q = \frac{d}{dt}(L_v) - L_q - K_q$$

$$\frac{d}{dt}(L''_{v'}) - L''_{q'} = \frac{d}{dt}(L'_{v'}) - L'_{q'} - K_q \ .$$

Hence, the definitive composite equations of motion are:

$$\boxed{\begin{aligned} \frac{d}{dt}(L_v) - L_q - K_q &= F_q \\ \frac{d}{dt}(L'_{v'}) - L'_{q'} - K_{q'} &= F'_{q'} \end{aligned}} \qquad (5.3)$$

These equations are especially interesting from the systems point of view. Suppose we consider q' as an *input* u, $x = (q, dq/dt)$ as a state. Then, the first equation of (4.3) takes the form:

$$\frac{d}{dt}(L_v(x)) - L_q(x) - K_q(x,u) - F_q(x) = 0 \ .$$

Similarly, the solutions of the first system may be considered as an input to the second.

Many of the examples of interconnected systems in the work of Kron and the RAAG group are constructed in this way. Of course, this is also a technique used extensively, to this day, in physics--particularly elementary particle physics and quantum field theory.

6. CONDITION THAT LAGRANGE'S EQUATIONS BE LINEAR, TIME-INVARIANT

Let $L(q,v)$ be a Lagrangian function. We want to investigate the conditions that the Lie group equations be linear. This means that there are relations of the following form (in matrix notation):

$$\frac{d}{dt}(L_v) - L_q = A\frac{d^2q}{dt^2} + B\frac{dq}{dt} + Cq \quad .$$

Hence,

$$L_{vv}\frac{d^2q}{dt^2} + L_{vq}\frac{dq}{dt} - L_q = A\frac{d^2q}{dt^2} + B\frac{dq}{dt} + Cq \quad .$$

Thus,

$$L_{vv} = A \, ,$$

$$L_{v_q} = B \, ,$$

$$L_q = Cq \quad .$$

A necessary condition is then that the *third* partial derivatives of L be zero, i.e., that:

L *is a quadratic function of the vector* (q,v).

7. QUADRATIC LAGRANGIANS AND LINEAR EQUATIONS

Adopt partitioned vector-matrix notation. Define L as follows:

$$L(q,v) = (q',v')\begin{pmatrix} A_{11} & A_{12} \\ A_{21} & A_{22} \end{pmatrix}\begin{pmatrix} q \\ v \end{pmatrix} + (b_1,b_2)\begin{pmatrix} q \\ v \end{pmatrix} \quad (7.1)$$

The Lagrangian differential operator

$$q \rightarrow \frac{d}{dt}\left(L_v\left(q,\frac{dq}{dt}\right)\right) - L_q \equiv \underline{L}(q) \quad . \quad (7.2)$$

can now be readily calculated (' denotes matrix transpose):

Engineering Systems 299

$$L_v = (0,1) A \begin{pmatrix} q \\ v \end{pmatrix} + (q',v')A \begin{pmatrix} 0 \\ v \end{pmatrix} + (b_1, b_2) \begin{pmatrix} 0 \\ 1 \end{pmatrix} .$$

Set:

$$v = \frac{dq}{dt}$$

$$a = \frac{dv}{dt}$$

$$\underline{L}(q) = (0,1)A \begin{pmatrix} v \\ a \end{pmatrix} + (v',a')A \begin{pmatrix} 0 \\ v \end{pmatrix} + (q',v')A \begin{pmatrix} 0 \\ a \end{pmatrix} - (1,0)A \begin{pmatrix} q \\ v \end{pmatrix}$$

$$- (q',v')A \begin{pmatrix} 1 \\ 0 \end{pmatrix} - (b_1, b_2) \begin{pmatrix} 1 \\ 0 \end{pmatrix}$$

This is a second order, linear, constant-coefficient differential operator in the vector-valued function of t, $t \to q(t)$.

8. F = MA AND THE EHRESMANN JET-CALCULUS

Let Q be a manifold, T a one-dimensional manifold used as a time interval, and let

$$J^1(T,Q), \; J^2(T,Q)$$

be the manifold of one- and two-jets of mappings $T \to Q$. Let

$$\pi: J^2(T,Q) \to J^1(T,Q)$$

be the "forgetting" map.

Here is how π is made into a *vector bundle*. It will be called the *acceleration bundle*. Assume, for the moment, that Q is a vector space. Let

$$\underline{q}:t \to q(t), \quad q':t \to q'(t)$$

be two curves in Q. Suppose that $t_0 \in T$ is a point of time such that

$$\partial q(t_0) = \partial q'(t_0) \quad .$$

(This means that $q(t) = q(t_0)$ and $q_t(t_0) = q'_t(t)$.) $\partial^2 q(t_0)$ and $\partial^2 q'(t_0)$ are two points of $J^2(T,Q)$ which lie in the *same fiber above* the point $\partial q(t_0) \in J^1(T,Q)$. Now, define the sum of these two points as

$$\partial^2 (q+q')(t_0) \quad .$$

This is the definition of "sum" that makes π into a vector bundle, i.e., defines a vector space structure on each fiber. One readily verifies that this vector bundle structure is *independent of the vector space structure on* Q *used to define it*.

To see what this has to do with physics, choose coordinates for $J^2(T,Q)$ labelled $(q,\dot{q},\ddot{q},t)$ so that:

$$q(\partial^2 q(t)) = q(t)$$

$$\dot{q}(\partial^2 q(t)) = dq(dt(t_0))$$

$$\ddot{q}(\partial^2 q(t)) = \frac{d^2 q}{dt^2}(t_0) \quad .$$

Similarly, define "Newtonian" coordinates for $J^1(T,Q)$. Then

$$\pi(q,\dot{q},\ddot{q},t) = (q,\dot{q},t)$$

$$\pi^*(q) = q$$

$$\pi^*(\dot{q}) = \dot{q}$$

$$\pi^*(t) = t \quad .$$

Of course, the vector bundle structure for π is such that:

$$(q,\dot{q},\ddot{q},t) + (q,\dot{q},\ddot{q}',t) = (q,\dot{q},\ddot{q}+\ddot{q}',t) \quad .$$

However, note that it makes no sense to add two elements of $J^2(T,Q)$ unless they have the same $(q,\dot{q},t)$-coordinates. Physically, q is the *position*, $\dot{q}$ the *velocity*, $\ddot{q}$ the *acceleration*.

Engineering Systems

This identification of "geometry" and "physics" suggests the natural formulation of *Newton's second law*.

Let $T^d(J^1(T,Q))$ be the *cotangent vector bundle* to $J^1(T,Q)$.

Definition. A *force law* is a cross-section map: $F: J^1(T,Q) \to T^d(J^1(T,Q))$ of this vector bundle.

A *mass operator* is a map

$$M: J^2(T,Q) \quad T^d(J^1(T,Q))$$

which is *linear* on the fibers of the vector bundle map π and the cotangent bundle.

If $\underline{q}: t \to q(t)$ is a curve in Q, its *acceleration* is the two-jet map:

$$t \to \partial^2 \underline{q}(t) \ .$$

It may be considered as a cross-section of the vector bundle π, pulled back under the map $\partial^1 \underline{q}$.

Definition. Given a mass operator M and a force law F, a *solution of Newton's equation* is a curve $\underline{q}: t \to q(t) \in Q$, which satisfies the following condition:

$$M(\partial^2 \underline{q}(t)) = F(\partial \underline{q}(t))$$

for all t.

9. A GENERAL DEFINITION OF MECHANICAL SYSTEMS IN TERMS OF EXTERIOR DIFFERENTIAL SYSTEMS

The Lagrangian point of view does lead to a certain exterior differential system where one-dimensional solution submanifolds are the trajectories of the mechanical system.

Let Q be a manifold, the configuration space of a mechanical system. Let T be a one-dimensional manifold, the time-interval manifold. Let $J^1(T,Q)$ be the one-jets of mappings: $T \to Q$. If (q^i), $1 \le i,j \le n$, are a coordinate system for Q, then (t, q^i, v^i) denote the natural coordinate system for $J^1(Q)$.

Set

$$\theta^i = dq^i - v^i dt .$$

They are the *contact* forms; the one-dimensional integral manifolds (on which $dt \neq 0$) are the tangent vectors to curves in Q. Let $\mathscr{C}$ be the $\mathscr{F}(J^1(Q))$-module of one-forms on $J^1(Q)$ generated by the θ^i. It is called the *contact system*. Then

$$d\theta^i = - dv^i \wedge dt .$$

Definition. A *mechanical system* with Q as configuration space, is defined by the following geometric data:

a) An exterior differential system $\mathscr{E}$ such that $\mathscr{E} \supset \mathscr{C}$.

b) A two-differential form ω such that $\omega \in \mathscr{E} \wedge \mathscr{C}$.

($\mathscr{E} \wedge \mathscr{C}$ denotes the set of one-forms which can be written as linear combinations of forms of the type: $\eta \wedge \theta$, with $\eta \in \mathscr{E}$, $\theta \in \mathscr{C}$.)

c) Locally, $\mathscr{E}$ is generated by the contact forms θ^i and an additional set of forms η_i such that

$$\omega = \eta_i \wedge \theta^i .$$

(Another way of putting this condition is to say that the tangent vector v annihilated by the one-forms in $\mathscr{C}$ are characteristic vectors of ω.) Thus, geometrically, a mechanical system is a *geometric structure* defined by the pair $(\mathscr{E}, \omega)$.

Let us briefly recall how the standard mechanics fits into this framework. For simplicity, let Q be one-dimensional with coordinate q. If

$$L(q,v)$$

Engineering Systems

is the Lagrangian, then

$$\theta = L\,dt + L_v(dq - v\,dt)$$

$$\omega = d\theta$$

$$= (dL_v - L_q\,dt) \wedge (dq - v\,dt) \quad .$$

We now investigate certain general geometric issues which suggest themselves. Suppose

$$\omega = \eta_i \wedge \theta^i \quad ,$$

when η_i are a set of one-forms, θ^i the contact forms defined by a specific coordinate system, labelled q^i, for Q. Let us look for conditions that ω be closed:

$$d\omega = d\eta_i \wedge \theta^i + \eta_i \wedge dv^i \wedge dt \quad .$$

Now, since

$$\theta^i = dq^i - v\,dt \quad ,$$

the one-forms

$$\theta^i, \quad dv^i, \quad dt$$

provide a basis for one-forms and $J^1(Q)$.

Suppose that:

$$\eta_i = a_{ij}\theta^j + b_{ij}dv^j + c_i\,dt \quad .$$

The one-dimensional integral curves of $\mathscr{E}$ (parameterized by t) are then solutions of the second order equations:

$$b_{ij}(q(t),v(t),t)\,\frac{d^2 q_i}{dt^2} + c_i(q(t),v(t),t) = 0$$

These are the *equations of motion* of the mechanical system.

Any system of second order equations which is linear in the second order derivatives can be written in this form. This is perhaps the ultimate version of

$$F = MA \;,$$

where the acceleration A is the second derivative, but "M" is more complicated, essentially functions b. (M is, in fact, a cross-section of a vector bundle over J^1. We will examine this point in more detail.

10. THE TWO-JET SPACE AS A VECTOR BUNDLE OVER THE ONE-JET BUNDLE

In mechanics, the second derivatives, i.e., the accelerations, often occur *linearly*, despite the fact that the lower derivatives are nonlinear. (Such equations are often called *quasilinear* in the classical language.)

Let T be an interval of real numbers and Q a manifold. Recall how the Ehresman jet spaces $J^r(T,Q)$ are defined. Let $\mathcal{M}(T,Q)$ be the space of smooth maps $\underline{q}: T \to Q$, i.e., the space of curves in Q.

Introduce an equivalence relation in

$$T \times \mathcal{M}(T,Q)$$

as follows:

$$(t,\underline{q}) \sim (t',\underline{q}')$$

if and only if

$$t = t'$$

$$\underline{q}(t) = \underline{q}'(t')$$

$$\frac{d\underline{q}}{dt}(t) = \frac{d\underline{q}'}{dt}(t)$$

$$\vdots$$

$$\frac{d^r\underline{q}}{dt^r}(t) = \frac{d^r\underline{q}}{dt^r}(t) \;.$$

Engineering Systems

In words, $\underline{q}$ and $\underline{q}'$ meet to the r-th order at the value $t = t'$.

$J^r(T,Q)$ is the quotient of $T \times \mathcal{M}(T,Q)$ by this equivalence relation. If $\underline{q} \in \mathcal{M}(T,Q)$, $\partial^2 \underline{q}$ is the map $T \to J^r(T,Q)$, which assigns to $t \in T$ the equivalence class to which $(t,\underline{q})$ belongs.

Since contact to r-th order implies contact to $(r-1)$-st order, there is a "forgetting highest derivatives" map

$$J^r(T,Q) \to J^{r-1}(T) .$$

This is readily seen to be a fiber space map, i.e., a submersion which is a local product. We aim to show that it also has a vector bundle structure. Let us restrict attention to the case

$$r = 2 ,$$

which, of course, is the most important for mechanics.

Let

$$\underline{q}: T \to Q$$

be a curve in Q. Let $\partial^1 \underline{q}$ be its one-jet. It is a map

$$T \to J^1(T,Q) .$$

Assign to (t,q) the tangent vector to the curve $t \to \partial^1 \underline{q}$. It is an element of

$$J^1(T,Q)_{\partial^1 \underline{q}(t)} ,$$

the tangent vector space $J^1(T,Q)$ at the point $\partial^1 \underline{q}(t)$. Call this

$$\phi(t,\underline{q}) .$$

This map ϕ is constant on the equivalence classes (for $r = s$) which define $J^2(T,Q)$, hence ϕ passes to the quotient to define a map

$$\underline{\phi}: J^2(T,Q) \to T(J^1(T,Q)) .$$

(The right hand side is the *tangent bundle* to the one-jet space.)

Now, consider the fiber of the "forgetting" map:

$$\pi: J^2(T,Q) \to J^1(T,Q) \ .$$

From the ϕ functions we see that two elements of $J^2(T,Q)$, which lie in the same fiber of π go over under ϕ to the same tangent space to $J^1(T,Q)$. In fact, it is readily seen that the map between the fiber of π and the tangent space is *one-one*. The map identifies the fiber with an affine subspace of the tangent space. (We will verify this in terms of local coordinates in a moment.) In general, an *affine subspace* of a vector space V is one of the form

$$\alpha = v_0 + S \ ,$$

where v_0 is a fixed element of V, and S is a linear subspace. This affine subspace is isomorphic to a vector space (if $v_0 \notin S$) in the following way:

Let β be the quotient map

$$V \to V/(v_0) \ ,$$

where $V/(v_0)$ is the quotient of V by the one-dimensional linear subspace spanned by v_0. β is then an isomorphism between the affine subspace α and the image of S in $V/(V_0)$.

Hence, each fiber of π is identified with an *affine space*, which provides π with a *vector bundle* structure.

It is not necessary to be so determidly "coordinate free". Let q be a coordinate system for Q. (Assume, for simplicity, that Q is one-dimensional.) Let $(q,\dot{q},t)$ and $(q,\dot{q},\ddot{q},t)$ be the usual "Newtonian" coordinates for $J(Q)$. π is then defined as follows:

$$\pi(q,\dot{q},\ddot{q},t) = (q,\dot{q},t)$$

Engineering Systems

Two points of J^2 go into the same point of J^1 if and only if their $(q,\dot{q},t)$-coordinates are the same.

Suppose now that $\underline{q}$ is a map $\pi \to Q$ given in the classical style by $t \to \underline{q}(t)$. $\partial^1\underline{q}$ is the map

$$t \to (q(t), dq/dt, t) \ .$$

Now, the basis of tangent vectors to $T(J^1(T,Q))$ corresponds to the coordinates $(q,\dot{q},t)$ may be labelled:

$$\partial_q \ , \ \partial_{\dot{q}} \ , \ \partial_t \ .$$

The tangent vector to the curve $\partial^1\underline{q}$ is then

$$\left(\frac{dq}{dt}\right)\partial_q + \left(\frac{d^2q}{dt^2}\right)\partial_{\dot{q}} + \partial_t$$

We see that the fiber of π goes over to the affine subspace

$$\left(\frac{dq}{dt}\right)\partial_q + \partial_t + \text{(multiples of } \partial_{\dot{q}})$$

Of course, old-fashioned tensor analysis also provides a definition of this vector bundle structure. When the coordinates (q) are changed, the *second derivatives*, modulo the first derivatives, change in a *linear* homogeneous way, which is the tip-off to the "vector bundle structure".

Finally, here is a direct way to define this vector bundle structure. Let $(t,\underline{q})$ be an element of $T \times \mathcal{M}(T,Q)$. Let $\mathcal{F}(Q)$ be the C^∞ real-valued functions on Q. Assign to $(t,\underline{q})$ the linear map: $\mathcal{F}(Q) \to R$ labelled $\delta^2(t,\underline{q})$ as follows:

$$\delta^2(t,\underline{q})(f) = d^2(dt^2(f(\underline{q}(t)))) \ .$$

The space of linear maps in $\mathcal{F}(Q)$ is a linear space. If $(t,\underline{q})$, $(t',\underline{q}')$ have the same second order of contact, then

$$\delta^2(t,q) = \delta^2(t',q') \ .$$

Hence, δ passes to the quotient to define a map

$$J^2(T,Q) \to L(\mathscr{F}(Q),R)$$

$(L(\mathscr{F}(Q),R) =$ vector space of known maps $\mathscr{F}(Q) \to R$.)

Theorem. The fibers of the forgetting map $J^2 \to J^1$ go over under this map to a subspace of $L(\mathscr{F}(Q),R)$.

Proof. Let us use coordinates (q). If $(t,\underline{q})$, with $\underline{q}: t \to \underline{q}(t)$, then $\delta^2(t,\underline{q})$ is the map

$$f \to f_{qq}\dot{q}^2 + f_q \ddot{q}$$

Q.E.D.

To obtain the vector space structure for the fiber of the forgetting map $J^2 \to J^1$, one has only to assign to each point of the fiber the quotient linear vector space corresponding to the affine subspace.

Remark. Here is the general setting. We assigned to each fiber $J^2(T,Q)$ an affine subspace

$$v_0 + S$$

of a vector space V. For two points in the same fiber, the point v_0 is the *same*. Thus, we can map the fiber into the quotient V/S, obtaining the vector bundle structure directly.

Chapter 19

NEWTON-LAGRANGE LINEAR SYSTEMS

1. NEWTON-LAGRANGE EQUATIONS AND VECTOR SPACES

Linear systems are, of course, very important in practical applications. In this chapter I will specialize the Lagrangian formalism to study the intersection of two classes of systems. I will start off with Lagrange's equations in R^n, then translate into a more abstract vector space notation.

Recall the previous work. Q is a manifold of dimension n, the configuration space of a mechanical system. Let (q^i), $1 \leq i, j \leq n$, a coordinate system for Q. In this chapter, we shall suppose that Q is a vector space and that the q^i are linear functions, i.e., a basis for Q^d, the dual space to Q. V denotes another copy of Q, the *velocity space*. L is a real valued function in $Q \times V$. v^i denote the functions q^i, relabelled with the notational change $Q \to V$.

$$L_{q^i}, \quad L_{v^i}, \quad L_{q^i q^j}, \quad \ldots$$

denote the partial derivatives with respect to these variables.

<u>Remark</u>. L_{q^i} can be defined as the map $Q \times V \to R$ such that

$$\frac{d}{dt} L(q + tq', v) = L_{q^i}(q + tq', b)\, q^i(q')$$

for $q, q' \in Q$.

(The summation convention is in use in this paper.) Further derivatives can be used in an analogous way.

Let us now define a first order differential operator Δ_L operator on curves in Q as follows:

$$\Delta_L(\underline{q}) = \left(\frac{d}{dt} \left(L_{v^i}\left(q(t), \frac{dq}{dt}\right)\right) - L_{q^i}\left(q(t), \frac{dq}{dt}\right)\right) q^i \qquad (1.1)$$

Then, for each curve

$$\underline{q}: t \to q(t)$$

in Q, $\Delta_L(\underline{q})$ is a curve in Q^d.

A *force law* is a map

$$F: Q \times F \to Q^d$$

Remark. This geometric-algebraic interpretation of the notion of "force" is a key link between modern differential geometry, mechanics, and control. It was first isolated, I believe, in *Geometry, Physics and Systems*. It was implicit in the work of the generation of tensor analysts.

Given a Lagrangian function L and a force law F, the *Newton-Lagrangian* equations are:

$$\Delta_L(\underline{q}) = F\left(q, \frac{dq}{dt}\right) \qquad (1.2)$$

to be solved for a curve $\underline{q}: t \to q(t)$.

2. QUADRATIC LAGRANGIAN

Q and V are copies of the same vector space. L is a real valued function in $Q \times V$. $Q \times V$ is itself a vector space, the direct sum of Q and V. One can now look for Lagrangian functions $L: Q \times V \to R$ which are quadratic or linear, i.e., L is of the form

$$L = L^2 + L^1, \qquad (2.1)$$

where L^2 is a quadratic function, L^1 is linear. Then L^2 and L^1 are of the following form:

$$L^2\left(\begin{pmatrix} q \\ v \end{pmatrix}\right) = (q, v) \begin{pmatrix} \alpha_{11} & \alpha_{21} \\ \alpha_{21} & \alpha_{22} \end{pmatrix} \begin{pmatrix} q \\ v \end{pmatrix}$$

$$= (q, v) \begin{pmatrix} \alpha_{11}q + \alpha_{12}v \\ \alpha_{21}q + \alpha_{22}v \end{pmatrix}$$

$$= \left\langle \begin{pmatrix} \alpha_{11}q + \alpha_{12}v \\ \alpha_{21}q + \alpha_{22}v \end{pmatrix}, \begin{pmatrix} q \\ v \end{pmatrix} \right\rangle$$

$$= \langle \alpha_{11}(q) + \alpha_{12}(v), q \rangle + \langle \alpha_{21}(q) + \alpha_{22}(v), v \rangle$$

$$\begin{aligned} &= \langle \alpha_{11}(q), q \rangle + \langle \alpha_{12}(v), q \rangle \\ &\quad + \langle \alpha_{21}(q), v \rangle + \langle \alpha_{22}(v), v \rangle \end{aligned} \quad (2.1)$$

$\alpha_{11}, \alpha_{21}, \alpha_{12}, \alpha_{22}$ are linear maps with the following domains and ranges:

$$\alpha_{11}: Q \to Q^d$$

$$\alpha_{12}: V \to Q^d$$

$$\alpha_{21}: Q \to V^d$$

$$\alpha_{22}: V \to V^d$$

$$L^1((q,v)) = \beta_1(q) + \beta_2(v),$$

$$\beta_1 \in Q^d; \quad \beta_2 \in V^d.$$

$$L_v^2 = \alpha_{12}^d(q) + \alpha_{21}(q) + 2\alpha_{22}(v) \quad (2.1)$$

$$L_v^1 = \beta_2 \quad (2.2)$$

$$L_q^2 = 2\alpha_{11}(q) + \alpha_{12}(v) + \alpha_{21}^d(v) \quad (2.3)$$

$$L_q^1 = \beta_1. \quad (2.4)$$

Thus, the Lagrange operator is:

$$\Delta_L = \alpha_{12}^d\left(\frac{dq}{dt}\right) + \alpha_{21}\left(\frac{dq}{dt}\right) + 2\alpha_{22}\left(\frac{d^2q}{dt^2}\right) - 2\alpha_{11}(q) \quad (2.5)$$

$$- \alpha_{12}\left(\frac{dq}{dt}\right) - \alpha_{21}\left(\frac{dq}{dt}\right)$$

A force law is of the form:

$$F(q,v) = \gamma_1(q) + \gamma_2(v) + u \quad (2.6)$$

with

$$C_1 : Q \to Q^d$$

$$C_2 : Q \to Q^d .$$

Finally then, we have:

<u>Theorem 2.1.</u> In this notation, the linear Newton-Lagrange equations are of the following form:

$$2\alpha_{22}\left(\frac{d^2q}{dt^2}\right) + (\alpha_{12}^d + \alpha_{21} - \alpha_{12} - \alpha_{21}^d)\left(\frac{dq}{dt}\right) - 2\alpha_{11}(q) - \beta_2 \qquad (2.7)$$

$$= \gamma_1(q) + \gamma_2 \frac{dq}{dt} + u$$

with $u \in Q^d$. (In the system theoretic interpretation, u will be the *input*.) In (2.7) $\alpha_{11}, \alpha_{22}, \alpha_{12}, \alpha_{21}$ are linear maps: $Q \to Q^d$,

$$\alpha_{11}^d, \alpha_{12}^d, \alpha_{21}^d, \alpha_{22}^d$$

are their duals,

$$\alpha_{11}^d = \alpha_{11} \qquad (2.8)$$

$$\alpha_{22}^d = \alpha_{22} , \qquad (2.9)$$

γ_1, γ are linear maps $Q \to Q^d$, β_2 and u are elements of Q^d.

<u>Remark.</u> If α_{22}^{-1} exists, note that (2.7) is readily put into input-state form, with u the input,

$$x = \left(q, \frac{dq}{dt}\right)$$

the state.

3. LINEAR NEWTON-LAGRANGE SYSTEMS FOR WHICH q IS A CYCLIC VECTOR

Equations (2.7) take the following form if the terms involving q vanish:

$$2\alpha_{22}\left(\frac{d^2q}{dt^2}\right) + (\alpha_{12}^d + \alpha_{21} - \alpha_{12} - \alpha_{21}^d)\left(\frac{dq}{dt}\right) - \gamma_2\left(\frac{dq}{dt}\right) = u \qquad (3.1)$$

Newton-Lagrange Systems

Remark. β_2 map be taken without loss in generality as zero.

Set

$$i = \frac{dq}{dt} . \tag{3.2}$$

(The notation should suggest "i = current", "q - charge", from electrical circuit theory.) Then,

$$2\alpha_{22}\left(\frac{di}{dt}\right) + (\alpha_{12}^d + \alpha_{21} - \alpha_{12} - \alpha_{22}^d)(i) - \gamma_2(i) = u \tag{3.3}$$

We can, of course, also write this as an exterior differential system:

$$2\alpha_{22}\, di + (\alpha_{12}^d + \alpha_{21} - \alpha_{12} - \alpha_{21})i\, dt = u\, dt \tag{3.4}$$

Assuming i = *current*, u = *voltage*, notice that the diverse *constitutive relations* of electrical circuit theory can be obtained by imposing constraints of the coefficient natures of (3.3). For example, *Ohm's Law*: $\alpha_{22} = 0$.

Chapter 20

THE GEOMETRIC NATURE OF "POWER" IN LAGRANGIAN MECHANICS ON MANIFOLDS

1. INTRODUCTION

The physical concepts of "energy" and "power" are fundamental in the practical applications of mechanics. However, these concepts have not had as much attention in the development of geometric system theory. A noted exception is the work of Jan Willems, who attempted to formulate the concepts of energy in a general systems framework in terms of an input-output description of systems. In this paper, I want to develop the *geometric Lagrangian* point of view. This gives a definite physical meaning to "energy", and enables one to formulate some basic concepts in terms of the geometry of manifolds.

2. LAGRANGIAN SYSTEMS

Let Q be an n-dimensional manifold. Let $\mathcal{F}(Q)$ be the (C^∞) real-valued functions on Q. (All geometric objects will be "smooth", i.e., of differentiability class C^∞, unless mentioned otherwise.)

$$\mathcal{D}(Q) = \{\mathcal{D}^j(Q), j=0,1,2,\ldots\}$$

denotes the graded, associative algebra of differential forms in Q. (The algebra operation is exterior multiplication, denoted as $\wedge$. $d: \mathcal{D}^j \to \mathcal{D}^{j+1}$ denotes the exterior derivative operation.) Suppose a coordinate system

$$q = (q^i), \qquad 1 \leq i,j \leq n$$

is fixed. Let X be the tangent vector bundle to Q, i.e. a point X is a pair (q,v) if a $q \in Q$ and a tangent vector $v \in Q_q$ to Q at q. Let (q_i, v^i) be the functions on $T(Q)$ such that:

$$q^i(q,v) = q^i(q)$$

$$v^i(q,v) = dq^i(v).$$

One readily proves that $(q,v) = (q^i,v)$ forms a coordinate system for the 2n-dimensional manifold $T(Q)$.

Another space involved in control theory of mechanical systems (where the "forces" are cotangent vectors) is the space Z defined as follows:

$$Z = \{(q,v,u): q \in Q,\ v \in Q_q,\ u \in Q_q^d\}\ .$$

Define a coordinate system

$$(q^i, v^j, u_k)$$

for Z as follows:

$$q^i(q,v,u) = q^i(q)$$

$$v^j(q,v,u) = dq^j(v)$$

$$u_k(q,v,u) = u\left(\frac{\partial}{\partial q^j}\right)$$

A *Lagrangian* L is an element of $\mathscr{F}(X)$. Set:

$$dL = L_{q^i}\, dq^i + L_{v^i}\, dv^i\ ,$$

i.e., the (L_{q^i}, L_{v^i}) are the partial derivatives of the function L with respect to the coordinates (q,v).

3. LAGRANGE'S EQUATIONS WITH FORCES AND CONTROL

Let Q be an n-dimensional manifold, as in Section 2, and let Z be the 3n-dimensional manifold constructed there from Q. (Z is the direct sum of the tangent and cotangent vector bundles to Q.) Let X be the tangent bundle to Q.

Consider Z as a vector bundle over X. The projection map is

$$(q,v,u) \to (q,v)\ . \tag{3.1}$$

Thus, each function (or differential form) on X can be considered as a function on Z, pulled back via the map (3.1). In control-system theory terms, the fibers u of this map are the "controls".

Suppose given a Lagrangian function, i.e., an element of $\mathscr{F}(X)$. Consider this function also on Z, pulled back via the projection $Z \to X$. Let (q,v,u) be the coordinates for Z constructed in Section 2. ((q,v) live on X.)

Let

$$F = F_i\, dq^i \tag{3.2}$$

be a one-differential form on X, of the indicated form. (F_i may be general elements of $\mathcal{F}(X)$, i.e., they may depend on v.)

Lagrange's equations with forces and controls are the following set of ordinary differential equations:

$$\frac{d}{dt}(L_{v^i}) - L_{q^i} = F_i + u_i \tag{3.3}$$

$$\frac{dx^i}{dt} = v^i \tag{3.4}$$

They determine a control system on the manifold Z_1 of the form:

$$f\left(x, \frac{dx}{dt}, u\right) = 0 \tag{3.5}$$

with $x = (q,v)$.

If they can be solved for dx/dt (which correspond to a *regular* or *non-singular* Lagrangian, in the classical terminology), then Equation (3.4) is a control system of the type that is now familiar, solved for the time-derivative of x. In the singular case, often the system 3.5 implies a control system of standard type, with additional *constraints* linking the "states" and "controls".

Our goal is now to adopt a definition of "energy" that is familiar from analytical mechanics, and to investigate its relation to the system equation (3.3).

4. ENERGY AND POWER

Keep the notation of Section 3. Set:

$$E = L_{v^i} v^i - L. \tag{4.1}$$

E, an element of $\mathcal{F}(X)$, is the (total) energy of the system. Let us calculate how E evolves along a solution $t \to (q(t),v(t),u(t))$ of (3.2)-(3.3):

$$\frac{d}{dt}(E) = \frac{d}{dt}(L_{v^i})v^i + L_{v^i}\frac{dv^i}{dt} - L_{q^i}v^i - L_v\frac{dv^i}{dt}$$

$$= (F_i + u_i)v^i \tag{4.2}$$

Set:

$$P = (F_i + u_i)v^i. \tag{4.3}$$

P is called the *power* of the system (3.2)-(3.3). Let us sum up as follows:

<u>Theorem 4.1</u>. Let E and P be the elements of $\mathcal{F}(Z)$ defined by (4.1) and (4.3). Then, along solutions

$$t \to (q(t), v(t), u(t))$$

of (3.3)-(3.4), we have

$$\frac{d}{dt}(E) = P. \qquad (4.4)$$

<u>Remark</u>. The novelty here is in the introduction of the *cotangent* vector variables (u) in the definition of power. It is readily seen that E and P are globally defined, i.e., independent of the local coordinates (q') for Q used to define them. (Note that the coordinates (v,u) are determined by the choice of q^i.)

5. THE SPACE OF ONE-JETS

My goal now is to put the classical material described in previous sections into a more coordinate-free form, in terms of naturally defined "geometric" structures. The formalism of "jets" of mappings, due to C. Ehresmann, seems well suited.

Let Q be a manifold, and let (q^i), $1 \leq i,j \leq n$, be a coordinate system of functions on Q. We shall construct a "prolonged" coordinate system on $J^1(R,Q)$, the space of one-jets of mappings: $R \to Q$.

Parameterize R, the "source" space, by the parameter τ. Denote an element of M(R,Q), the space of mappings $R \to Q$, by $Q: \tau \to q(\tau)$. Recall that $J^1(R,Q)$ is the quotient of $R \times M(R,Q)$ by the following equivalence relation, namely:

(t,q) is equivalent to (t',q') if and only if
t = t', and q,q' have the same tangent vector as t.

Define real-valued functions

$$q^i, v^i, t: R \times M(R,Q) \to R$$

by the following formulas:

$$q^i(\tau, q) = q^i(q(\tau)),$$

$$v^i(\tau, q) = \frac{d}{d}(q^i(q(\tau)))$$

Power

$$t(\tau,q) = \tau .$$

Notice that these functions pass to the quotient to define functions on $J^1(R,Q)$; we shall use the same notation for these functions. One now proves readily that:

The functions $(q^i, \dot{q}^i, t)$ define a coordinate system for $J^1(R,Q)$.

In terms of this natural "prolonged" coordinate system for $J^1(R,Q)$, let us now define the following differential forms:

$$\rho^i = dq^i - v^i \, dt . \tag{5.1}$$

A Lagrangian function L is now interpreted as a real-valued function of $\mathcal{F}(J^1(R,Q))$. Set:

$$\omega = (d(L_{v^i}) - L_{q^i} \, dt) \wedge \rho^i \tag{5.2}$$

$$F = F_i \rho^i . \tag{5.3}$$

We can, of course, identify $J^1(R,Q)$ with the space of triples

$$(t,q,v) ,$$

with:

$t \in T$; $q \in Q$; $v \in Q_q$, the tangent space of Q at q.

Let Z be the space of quadruples

$$(t,q,v,u) ,$$

with:

$$u \in Q_q^d .$$

The map

$$(t,q,v,u) \to (t,q,v)$$

defines Z as a *vector bundle* over $J^1(R,Q)$. The fibers are the space of cotangent vectors to Q.

Let $u^i \in \mathcal{F}(Q)$ be the functions such that:

$$u_i(t,q,v,u) = u\left(\frac{\partial}{\partial q^i}\right) \tag{5.4}$$

Set:

$$\alpha = u_i \rho^i \:. \tag{5.5}$$

This one-form on Z is obviously invariantly defined, independently of the coordinate system.

Theorem 5.1. The curves

$$t \to (t, q(t), v(t), u(t))$$

in Z that are solutions of Lagrange's equations with internal and external forces are the integral curves of vector fields V on Z that satisfy the following relations:

$$V \lrcorner \omega = F + \alpha \tag{5.6}$$

$$\rho^i(V) = 0 \tag{5.7}$$

$$dt(V) = 1 \tag{5.8}$$

Now, we can put the energy function E into this framework. Set:

$$\theta = L dt + L_{v^i} \rho^i \:. \tag{5.9}$$

One proves readily that

$$d\theta = \omega \tag{5.10}$$

Then, we have:

$$E = \partial_t \lrcorner \theta \tag{5.11}$$

(∂_t is the vector field $\partial/\partial t$, i.e., the infinitesimal generator of time-translation.)

Theorem 5.2. E, defined by formula (5.11), satisfies:

$$dE = \partial_t \lrcorner \omega \tag{5.12}$$

i.e., "energy" is "*conjugate*" to "time".

Proof.

$$dE = -d(\partial_t \lrcorner \theta)$$

$$= -\mathscr{L}_{\partial_t}(\theta) + \partial_t \lrcorner \omega \:.$$

Power 321

But, $\mathcal{L}_{\partial_t}(\theta) = 0$, since the Lagrangian L is invariant under time-translation.

Now, we can define "power" as the Lie derivative of the function E with respect to the vector field V defined by relations (5.6)-(5.8)

$$P \equiv \mathcal{L}_V(E)$$

$$= \quad , \text{ using (5.11)},$$

$$- \mathcal{L}_V(\partial_t \,\lrcorner\, \theta)$$

$$= - \partial_t \,\lrcorner\, \mathcal{L}_V(\theta)$$

$$= - \partial_t \,\lrcorner\, (V \,\lrcorner\, d\theta) - d(\theta(V))$$

$$= - \partial_t \,\lrcorner\, (F + \alpha) + \partial_t(\theta(V))$$

$$= - (F + \alpha)(\partial_t) \quad . \tag{5.9}$$

Theorem 5.3. Formula (5.9) exhibits "power" as a function intrinsically defined on the vector bundle Z over $J^1(R,Q)$, which is the pull-back of the cotangent bundle to Q under the projection map $J^1(R,Q) \to Q$.

Chapter 21

THE RLC EQUATION IN GEOMETRIC FORM

1. INTRODUCTION

Underlying the work of Kron and the Kondo group are analogies at the geometric level between *electrical circuit theory* and *Lagrangian* mechanics. In this chapter I will make some relations of this type more explicit.

2. THE DAMPED-HARMONIC OSCILLATOR-RLC EQUATIONS IN COORDINATE-FREE FRAMES

Let Q be a real finite dimensional vector space. (It is possible to generalize much of this material to the infinite dimensional situation.) Let Q^d be its dual space. Let $\underline{Q}$ and $\underline{Q}^d$ be the space of C^∞-maps, $R \to Q$ and $R \to Q^d$, i.e., geometrically the space of curves $\underline{q}: t \to q(t), \underline{q}^d, t \to \underline{q}^d(t)$ in these vector spaces. Let

$$\Delta: \underline{Q} \to \underline{Q}^d$$

be a linear differential operator of the form:

$$\Delta(\underline{q}) = C\underline{q}_{tt} + R\underline{q}_t + L\underline{q} \ . \tag{2.1}$$

(Subscripts denote time-derivatives.) C, R and L are linear maps: $Q \to Q^d$. Further, they are *symmetric*, i.e., satisfy the following identity:

$$\langle Cq_1, q_2 \rangle = \langle Cq_2, q_1 \rangle \tag{2.2}$$

for $q_1, q_2 \in Q$.

Let $\underline{v}: t \to v(t)$ be a curve in $\underline{Q}^d$. The differential equations

$$\Delta \underline{q} = \underline{v} \ , \tag{2.3}$$

or

$$C\underline{q}_{tt} + R\underline{q}_t + L\underline{q} = \underline{v} \tag{2.4}$$

represent physically the *damped harmonic oscillator with* $\underline{v}$ *as external force*, or, in current theory, a sort of *generalized RLC-circuit*. Of course, from the system theoretic point of view, (2.4) can be viewed as an *input system*, with *input* $\underline{v}$. Of course, the proper choice of *state* is not so evident -- and, in fact, will depend on the algebraic properties of the type (C, R, L) of linear maps.

3. POWER

Suppose

$$(q, v)$$

is a solution of equation (2.4), i.e., a C^∞ curve

$$t \to (q(t), v(t))$$

in $Q \times Q^d$, which satisfies Equation (2.4). Set:

$$P(q,v)(t) = \langle q_t(t), v(t) \rangle \tag{3.1}$$

$P(q,v)$ is thus a real-valued function of t, called the *power* function.

Let us use Equation (2.4) to derive certain relations that it satisfies:

$$\langle q_t, v \rangle = \langle q_t, Cq_{tt} + Rq_t + Lq \rangle$$

$$= \tfrac{1}{2} \langle Cq_t, q_t \rangle_t + \tfrac{1}{2} \langle Lq, q \rangle_t + \langle q_t, Rq_t \rangle \tag{3.2}$$

Motivated by the physics, let us define *kinetic*, *potential*, *total energy*, and *dissipation*.

$$KE(q,v) = \tfrac{1}{2} \langle Cq_t, q_t \rangle \tag{3.3}$$

$$PE(q,v) = \tfrac{1}{2} \langle Lq, q \rangle \tag{3.4}$$

$$E(q,v) = KE(q,v) + PE(q,v) \tag{3.5}$$

$$D(q,v) = \langle q_t, Rq_t \rangle \tag{3.6}$$

Thus, we have:

$$P(q,v) = \frac{d}{dt}(E(q,v)) + D(q,v) \ .$$

This is the basic "conservation of energy" relation.

4. THE "TRANSFER FUNCTION" OF THE RLC-SYSTEM AS A RATIONAL MAP FROM $P_1(\mathbb{C})$ TO A GRASSMANNIAN

Let i denote a vector of the underlying vector space Q. (Geometrically, i is really in the "tangent bundle to Q", which we can identify with Q, since Q is "flat". Let s be a complex variable. Let

$$T(s)(i) = sCi + sRi + \frac{1}{s}L \quad . \tag{4.1}$$

Remark. In terms of contemporary analysis, (4.1) should be considered as the "symbol of a pseudo-differential operator."

Let us associate with the RLC system a mapping of the Riemann sphere (the complex plane with the point at infinity) or *the one-dimensional complex projective space* $P_1(\mathbb{C})$. Set

$$\tau(s) = \{(i,v) : T(s)i = v, \ i \in Q, \ v \in Q^d\} \tag{4.2}$$

Suppose

$$n = \dim Q \quad .$$

Set:

$$G^n(Q \oplus Q^d) = \text{Grassmann manifold of n-dimensional linear subspace of the vector space } Q \oplus Q^d.$$

τ is thus a *rational mapping* of

$$P_1(\mathbb{C}) \to G^n(Q \oplus Q^d) \quad .$$

5. STATE SPACE FORM FOR RLC SYSTEMS IN THE NON-SINGULAR CASE

It is folklore in electrical engineering that state variables for circuits can generically be chosen as *inductor currents* and *capacitor voltages*. Let us see that this means for the sort of RLC-system described in previous sections:

$$Cq_{tt} + Rq_t + Lq = v \quad . \tag{5.1}$$

Definition. The RLC system is *non-singular* if C is an isomorphisms between the vector spaces Q and Q^d.

In this non-singular case, map a solution

$$t \to (q(t), v(t)) \equiv \underline{q}, \underline{v})$$

into a curve in $Q^d \times Q \times Q^d$ by the formula:

$$(\underline{q}, \underline{v}) = (C\underline{q}, \underline{q}_t, v) \tag{5.2}$$

Set:

$$\underline{x} = (C^{-1}\underline{q}, \underline{q}_t) \tag{5.3}$$

$$x = \begin{pmatrix} C^{-1}\underline{q} \\ \underline{q}_t \end{pmatrix}$$

$$X = Q^d \times Q$$

i.e., X is the direct sum of the vector space Q^d and its dual. More geometrically, X is the *cotangent bundle* to Q^d.

Differentiate both sides of (5.3):

$$\underline{x}_t = (C\underline{q}_t, \underline{q}_{tt})$$

$$= (C\underline{q}_t, -C^{-1}R\underline{q}_t - C^{-1}L\underline{q} + C^{-1}v) \quad . \tag{5.4}$$

Let us write this as an input system, with x as state vector:

$$x_t = Ax + Bv \quad , \tag{5.5}$$

A is to be a linear map

$$Q^d \times Q \to Q^d \times Q^d \quad ,$$

of the form:

$$A(q^d, q) = (\alpha q^d + \beta q, \gamma q^d + \delta q^d) \tag{5.6}$$

Then,

$$A\underline{x} = A(C\underline{q}, \underline{q}_t)$$

$$= (\alpha C\underline{q} + \beta \underline{q}_t, \gamma C\underline{q} + \delta \underline{q}_t) \quad . \tag{5.7}$$

Comparing (5.4) and (5.7), we have:

RLC Equations

$$\alpha = 0 \qquad (5.8)$$

$$\beta = C \qquad (5.9)$$

$$\gamma = -C^{-1}LC^{-1} \qquad (5.10)$$

$$\delta = -C^{-1}R \qquad (5.11)$$

$$Bv = (0, C^{-1}v) \qquad (5.12)$$

Let us sum up as follows:

<u>Theorem 5.1</u>. Let A be the linear map:

$$Q^d \times Q \to Q^d \times Q$$

defined as follows:

$$A(q^d, q) = (Cq, -C^{-1}LC^{-1}q^d - C^{-1}Rq) \qquad (5.13)$$

Let $B: Q^d \to (Q^d, Q)$ be:

$$B(v) = (0, C^{-1}v) \qquad (5.14)$$

Then, the RLC equations are transformed, under the map

$$\phi(\underline{q}, \underline{v}) = (\underline{x}, \underline{v})$$

$$\underline{x} = (C\underline{q}, \underline{q}_t) \in Q^d \times Q$$

into the state space equations:

$$\underline{x}_t = A\underline{x} + B\underline{v} .$$

6. RLC EQUATIONS COUPLED TO RIEMANNIAN METRICS -- THE KRON-HOFFMAN-KONDO EQUATIONS

Now, let Q be a vector space, with a linear coordinate system

$$(q^i) , \qquad 1 \leq i, j \leq n .$$

Let X be a manifold with coordinates

$$(x^a) , \qquad 1 \leq a, b \leq m .$$

Let $(L_{ij}(x))$, $(R_{ij}(x))$, $(C_{ij}(x))$ be symmetric, $n \times n$ matrices which are functions of $x \in M$. Define a Lagrangian for the manifold $X \times Q$ as follows:

$$K = \frac{1}{2} C_{ij}(x) \dot{q}^i \dot{q}^j + \frac{1}{2} L_{ij}(x) q^i q^j + \frac{1}{2} g_{ab}(x) \dot{x}^a \dot{x}^b , \qquad (6.1)$$

where

$$ds^2 = g_{ab} dx^a dx^b$$

is a *Riemannian metric* for X. Similarly, let

$$R = R_{ij} \dot{q}^i \dot{q}^j . \qquad (6.2)$$

Let us compute the Lagrange-Rayleigh equations with these choices for Lagrangian and Rayleighian

$$K_{\dot{q}^i} = C_{ij} \dot{q}^j$$

$$K_{q^i} = L_{ij} q^j$$

$$\frac{d}{dt}\left(K_{\dot{q}^i}\right) = C_{ij,x^a} \dot{x}^a \dot{q}^j + C_{ij} q^j_{tt}$$

$$L_{\dot{x}^a} = g_{ab} \dot{x}^b$$

$$\frac{d}{dt}\left(L_{\dot{x}^a}\right) = g_{ab,x^c} \dot{x}^c \dot{x}^b + g_{ab} x^b_{tt}$$

$$K_{x^a} = g_{bc,x^a} \dot{x}^b \dot{x}^c + \frac{1}{2} C_{ij,x^a} \dot{q}^i \dot{q}^j + \frac{1}{2} L_{ij,x^a} q^i q^j$$

Let us consider the Lagrange-Rayleigh equations with exterior forces only for the i-j indices:

$$u^i = \frac{d}{dt}\left(L_{\dot{q}^i}\right) - L_{q^i}$$

$$= C_{ij,x^a} \dot{x}^a \dot{q}^j + C_{ij} q^j_{tt} - L_{ij} q^j$$

$$0 = \frac{d}{dt}\left(L_{\dot{x}^a}\right) - L_{x^a}$$

$$= g_{ab,x^c} x_t^c x_t^b + g_{ab} x_{tt}^b - g_{bc,x^a} x_t^b x_t^c$$

$$-\frac{1}{2} C_{ij,x^a} q_t^i q_t^j - \frac{1}{2} L_{ij,x^a} q^i q^j \qquad (6.3)$$

(6.2)-(6.3) then constitute a set of (n+m)-ordinary differential equations in (n+m) unknowns,

$$t \to (q^i(t), x^a(t))$$

They are the equations considered in the work of Kron, Hoffman, and the RAAG group.

One can now put them into more (globally) geometric coordinate-free form. Notice that they are composed of three parts:

I. Linear equations in q, with x as parameters:

$$u = (C) x_t q_t + C q_{tt} - Lq$$

II. Geodesic-type equations in x:

$$(\partial\Gamma) x_t x_t + g x_{tt} \quad ,$$

where Γ are "Christoffel symbols" constructed from the g's.

III. Nonlinear interaction terms between x and q:

$$(\partial C) q_t q_t + (\partial L) qq$$

PART VI

GEOMETRIC STRUCTURE IN FIELD THEORY

The relation between quantum mechanics, symplectic structures, Poisson bracket/Poisson manifolds/cosymplectic structures,... is more or less well understood in the particle mechanics/finite number of degrees of freedom case. However, the extension to *field theories* remains a great challenge, which has already motivated much work in my previous books.

My own first work, which I believe even has certain claim to priority, is in the 1960's in my book, *Lie Algebras and Quantum Mechanics*, *Vector Bundles in Mathematical Physics*, and papers I wrote at the time in *Physical Review* and *Journal of Mathematical Physics* [4]. (Numbers refer to references at the end of Chapter 22.) Another historical phenomenon is the work in the elementary particle physicists' work in the 1960's on the so called σ-model. (Identical to what Eels and Sampson called, in earlier work, *harmonic maps* between Riemannian manifolds.) These models are strikingly involved in the integrability game!

The nature of Chapter 18 is the beginning of a return to this material to understand its geometric meaning in terms of the work of Lie and Cartan, and their modern followers: Goldschmidt and Sternberg [13] have emphasized the role of the *affine structure* naturally defined on the jet spaces. There is an associated vector bundle, and then a dual bundle. It is this that I call the *cojet bundles*. In later work (e.g., Volume 24 or 25) I plan to use this chapter as a take-off point and to introduce more extensive material on quantum field theory, elementary particle physics, integrable field theories, ...

Chapter 22

THE HAMILTON-VOLTERRA FIELD THEORY EQUATIONS AND THE COJET BUNDLES

1. INTRODUCTION

In previous work [1], I have pointed out the relevance of Lie's work [2] on "function groups" to topics of current interest in the theory of "integrable" systems. In Refs. 3-5 I have remarked that certain Lagrangian field theories (e.g., the σ or *Sugawara* model [6], known to mathematicians as the *harmonic maps*) could be considered in an analogous spirit. My aim now is to search for the generalization of the "function group" notion to dynamical systems with an infinite number of degrees of freedom.

Now, in a modern setting, the "function groups" can be described as follows. Start off with a manifold Q, with $T(Q)$ the tangent bundle. Consider the dual vector bundle $T^d(Q)$, i.e., the *cotangent bundle*. It carries a natural symplectic structure, which generates (via Poisson bracket) a Lie algebra structure on $\mathscr{F}(T^d(Q))$, the C^∞ real valued functions on $T^d(Q)$. The "function groups" are certain types of Lie subalgebras of this Poisson bracket structure.

In Ref. 3, I have indicated how the Poisson bracket Lie algebra structure could be extended to higher degree differential forms, and this serves as a geometric-algebraic setting for field theories. In 1890 [2] Volterra indicated how the Hamilton-Jacobi formalism could be extended to field theories. In Ref. 3, pp. 178-189, I have briefly shown how the Volterra formalism could be described by differential forms. My aim in this paper is to develop further the geometric foundations of the Volterra theory, with the ultimate aim of seeing how the "integrable" field theories (Korteweg-de Vries, Sine-Gordon,...) can be described in its terms, and to then search systematically and "rationally" for the physical systems which are then "integrable" in the Lie-function groups sense. (However, these applications will be dealt with in a later work.)

The geometric foundation of field theories is closely linked to Ehresmann's theory of *jets* of mappings [8]. Given two finite dimensional, C^∞, paracompact manifolds X and Y, they form a sequence

$$J^0(X,Y) = X \times Y, \; J^1(X,Y), \; J^2(X,Y), \ldots$$

of manifolds of increasing dimension, with natural projection maps $J^{r+1} \to J^r$. Hubert Goldschmidt [13] has pointed out that these maps can be given a natural affine structure, i.e., for each r, there is a vector space V^r, a Lie group G^r which acts on V^r in an affine way, and a principal

G^r-bundle over J^r, *such that the map* $J^{r+1} \to J^r$ *can be identified with* the bundle of fiber V^r associated with the action of G^r on V^r. Now, *any* affine action of G^r gives rise to a *linear* action of a quotient of actions *linearly* on a vector space isomorphic to V^r. Thus, there is *also a vector bundle*

$$E^{r+1} \to J^r$$

associated with the projection map $J^{r+1} \to J^r$. We obtain in this way a sequence of vector bundles

$$E^1 \to J^0(X,Y)$$

$$E^2 \to J^1(X,Y)$$
$$\vdots$$

The work in this chapter is, on the geometric side, related to previous work by Goldschmidt and Sternberg [13] and Shadwick [14]. They have done a much more systematic and complete job of describing the complicated jet structures and how the "Hamilton-Poincaré-Cartan" forms are defined and manipulated. My main goal here is to prepare the way for later work on the "σ-model" (or "harmonic maps") of the physicists [6]. In previous work [5], I have shown that they have an algebraic structure (which the physicists call the "current algebra" structure), which seems to be an indicative "integrability". My belief is that this structure is a generalization of the "functions groups" of Lie, i.e., what is involved is a generalization of Lie "function groups" to the algebraic structure defined on higher degree differential forms on manifolds, particularly the jet spaces of mappings onto Lie groups and other Riemannian manifolds. It is particularly noteworthy that the "current algebra" structures defined, in the 1960's [6], on the σ-model Lagrangians reduce, in the case the underlying domain manifold is two, i.e., one space, one time dimension field theories, to the *Kac-Moody algebras*, which have been extensively used in the recent integrability work. It is again unfortunate that the low state of scholarly awareness in the mathematical physics community has obscured this fundamental fact! A main point of this paper is that the *dual bundles* are fundamental for the geometrization of the Volterra formalism: I will call them the *cojet bundles*.

Note how this works in the special case of particle mechanics, i.e., $X = R$. $J^1(R,Y)$ can be identified with

$$R \times T(Y) ,$$

where $T(Y)$ is the tangent vector bundle to Y. The projection

$$J^1(R,Y) \to J^0(R,Y)$$

is, in this case, just the projection map:

$$R \times T(Y) \to R \times Y \qquad (1.2)$$

$$(t,v) \to (t,y)$$

for $y \in Y$, $v \in Y_y$.

The fiber of the map (1.2) above the point $(t,y) \in R \times Y \equiv J^0(R,Y)$ is the tangent space Y_y to Y above the point y, whence the vector space structure on each fiber required to call it a "vector bundle". The dual vector bundle is then:

$$R \times T^d(Y)$$

where $T^d(Y)$ is the cotangent vector bundle to Y, i.e., the dual vector bundle to the tangent bundle $T(Y)$.

As another preliminary matter, let us recall Cartan's formulation [9] of particle mechanics in terms of what he called *Cauchy characteristics* of two-differential forms. Let M be a finite dimensional, C^∞, paracompact manifold, with ω a two-differential form on M of constant rank on M. Then, there is (at least locally) an integer n, and $2n$ independent one-differential forms on M:

$$\alpha_1, \ldots, \alpha_n; \beta_1, \ldots, \beta_n$$

such that:

$$\omega = \alpha_1 \wedge \beta_1 + \cdots + \alpha_n \wedge \beta_n . \qquad (1.3)$$

The Pfaffian system

$$\alpha_1 = 0 = \cdots = \beta_1 = \cdots = \beta_n \qquad (1.4)$$

is intrinsically attached to ω; its integral curves are said to be *characteristic curves of* ω.

If the following condition is satisfied:

$$d\omega = 0 , \tag{1.5}$$

then the Pfaffian system (1.4) is *completely integrable* in the Frobenius sense. Its integral curves are then said to be *Cauchy characteristic curves of* ω. The characteristic curves passing through each point p of M then fill up a manifold of dimension

$$\dim M - 2n \tag{1.6}$$

called the *leaf* of the *Cauchy characteristic foliation*. We denote the foliation by

$$C(\omega) . \tag{1.7}$$

The quotient space

$$C(\omega) \setminus M ,$$

i.e., the space whose points are the leaves of the foliation (1.4), is not necessarily a manifold, in the sense that it can be provided with a natural Hausdorff topology compatible with the charts provided by the "flat" coordinate systems. If it does have this property, the foliation is said to be *regular*, and the quotient map

$$\pi: M \to C(\omega) \setminus M$$

is a *submersion* map, i.e., the induced map

$$\pi_*: T(M) \to T(C(\omega) \setminus M)$$

on tangent vectors is onto. (And the dual map on covectors and differential forms is one-one.)

If the foliation is regular, $C(\omega) \setminus M$ is a manifold of dimension $2n$, and carries a two-form $\bar{\omega}$ such that

$$\pi^*(\bar{\omega}) = \omega .$$

$C(\omega) \setminus M$ is a *symplectic manifold*.

If the Cauchy characteristic foliation is not regular, the situation is in a more confused state. The quotient space $C(\omega) \setminus M$ carries a *pseudogroup structure*, in the sense of Ehresmann. This pseudogroup is locally isomorphic with the pseudogroup of local canonical transformations acting on R^{2n}, preserving the usual symplectic structure.

A natural invariant of this structure is the *foliation holomony group*, in the sense of Ehresmann. (In [1], I have sketched a formalism which the holomony group could be obtained in a natural way as a "linear isotropy subgroup" of a *pseudogroup* associated with the foliation.)

Turning now to field theory, one way of thinking about their geometric structure is that they involve replacing ω by a higher degree differential form. Suppose that ω is such a form of degree m. (m-1 would then be the number of space-time variables.) Consider now decompositions of ω of the form (1.3), where the $\alpha_1,\ldots,\alpha_n$ are of degree one, but the $\beta_1,\ldots,\beta_n$ are of degree (m-1). Consider the maximal integral submanifolds of the exterior differential system (1.4) (in the sense of Cartan [10]) as the *solutions of the field equations*. In previous work [3,11,13], I have shown how field theory problems arising from variational principles could be written in this way. My aim in this paper is to develop further the brief treatment given in Ref. 3 of the Volterra equation, in terms of Cartanian formulation, and to pin down the space M in terms of the Ehresmann jet bundles.

2. THE VOLTERRA FORMALISM IN LOCAL COORDINATES

Let us begin with a short description of the Volterra formalism in the classical spirit. Choose indices as follows:

$$0 \leq \mu,\nu \leq n-1$$
$$1 \leq a,b \leq m \; . \tag{2.1}$$

Suppose M is a manifold (assumed paracompact and C^∞) of dimension $n + m + nm$, with a coordinate system of functions that we label as follows:

$$(x^\mu, \phi^a, \pi^\mu_a) \equiv (x, \phi, \pi) \; . \tag{2.2}$$

Thus, we are supposing that M is a product:

$$M = X \times Y \times Z \tag{2.3}$$

with manifolds X, Y, Z, which have coordinates labelled x^μ, ϕ^a, π^μ_a, respectively. Let $H: M \to R$ be a real-valued C^∞ function on M. Let

$$H_\mu = \frac{\partial H}{\partial x^\mu}$$

$$H_a = \frac{\partial H}{\partial \phi^a} \tag{2.4}$$

$$H^a_\mu = \frac{\partial H}{\partial \pi^\mu_a}$$

and so on.

Definition. The *Hamilton-Volterra equations* for maps $\psi: X \to Y$, with Hamiltonian H, are the following systems of partial differential equations with

$$\psi^*(\phi^a) \equiv \phi^a(x) \equiv \phi(x)$$
$$\psi^*(\pi^\mu_a) = \pi^\mu_a(x) \tag{2.5}$$

$$\frac{\partial \phi^a}{\partial x^\mu} = H^a_\mu(x, \phi(x), \pi(x)) \equiv \frac{\partial H}{\partial \pi^\mu_a}$$
$$\frac{\partial \pi^\mu_a}{\partial x^\mu} = -H_a(x, \phi(x), \pi(x)) \equiv -\frac{\partial H}{\partial \phi^a} \tag{2.6}$$

In case $n = 1$, Equations (2.6) reduce to a system of ordinary differential equations with "x^0" the "independent variable". These equations take the familiar Hamiltonian form of classical analytical mechanics if x^0 is relabelled as "t".

3. THE HAMILTON-VOLTERRA EQUATION IN TERMS OF DIFFERENTIAL FORMS AND EXTERIOR DIFFERENTIAL SYSTEMS

I will now review the material in Ref. 3, pp. 178-188, showing how the Hamilton-Volterra Equations (2.6) can be described in terms of differential forms; this is in parallel with Cartan's description [9] of the Hamilton equations as "Cauchy characteristic" equations associated with differential forms. See also Ref. 12.

Keep the notation of Section 2. Introduce the following differential forms and vector fields on the manifold M

$$\partial_\mu = \frac{\partial}{\partial x^\mu} \tag{3.1}$$

$$dx = dx_0 \wedge \ldots \wedge dx_{n-1} \tag{3.2}$$

$$dx_\mu = \partial_\mu \,\lrcorner\, dx \tag{3.3}$$

$$\alpha^a = d\phi^a - H^a_\mu dx^\mu \tag{3.4}$$

$$\beta_a = d(\pi^\mu_a dx_\mu + H_a dx) \tag{3.5}$$

$$\theta = -\pi^\mu_a d\phi^a \wedge dx_\mu + H\, dx \tag{3.6}$$

Hamilton-Volterra and Cojet Bundles

Theorem 3.1.

$$d\theta = \alpha_a \wedge \beta^a \tag{3.7}$$

Proof. Apply the exterior differentiation operation "d" to (3.6):

$$d\theta = -d\pi_a^\mu \wedge dy^a \wedge dx_\mu + (H_a d\phi^a + H_\mu^a d\pi_a^\mu) \wedge dx \tag{3.8}$$

Use (3.4)-(3.5):

$$\alpha_a \wedge \beta^a = (d\phi^a - H_\mu^a dx^\mu) \wedge (d\pi_a^\mu \wedge dx_\mu + H_a dx)$$

$$= d\phi^a \wedge d\pi_a^\mu \wedge dx_\mu + d\phi^a \wedge H_a dx - H_\mu^a dx^\mu \wedge d\pi_a^\nu \wedge dx_\nu \tag{3.9}$$

Now,

$$0 = \partial_\nu \lrcorner (dx^\mu \wedge dx)$$

$$(\partial_\nu \lrcorner dx^\mu) \wedge dx - dx^\mu \wedge dx_\nu$$

whence:

$$dx^\mu \wedge dx_\nu = \delta_\nu^\mu dx \tag{3.10}$$

Combine (3.9) and (3.10):

$$\alpha_a \wedge \beta^a = d\phi^a \wedge d\pi_a^\mu \wedge dx_\nu + H_a d\phi^a \wedge dx + H_\mu^a d\pi_a^\mu \wedge dx \tag{3.11}$$

The right hand sides of (3.11) and (3.8) are equal, whence formula (3.7) and Theorem 3.1. Q.E.D.

Theorem 3.2. Let $\mathscr{E}(H)$ be the exterior differential system on M generated by the forms α_a, β^a, i.e., $\mathscr{E}(H)$ is the smallest differential ideal of $\mathscr{D}(\mathscr{E})$ containing the α_a and β^a. Then, a map

$$\psi: x \to (x, y(x), z(x))$$

of $X \to M$ is a solution of the Hamilton-Volterra Equations (2.6) if and only if ϕ is a solution submanifold of the exterior differential system $\mathscr{E}(H)$, i.e.,

$$\phi^*(\alpha_a) = \phi^*(\beta^a) = 0 .$$

Proof. This should be clear on comparing formulas (3.4)-(3.5) and (2.6).

4. THE HAMILTON-VOLTERRA EQUATIONS AND THE ONE-COJET BUNDLE

Theorem 3.2 establishes that the Hamilton-Volterra Equations (2.6) have a form independent of the coordinates $(x^\mu, \phi^a, \pi^\mu_a)$ used to define them. However, we can go further, and identify the manifold M on which they "live", defined *intrinsically* in terms of the manifolds X and Y.

Recall how the space of one-jets, $J^1(X,Y)$, is defined. Start with ordered pairs

$$(x, \psi) \qquad (4.1)$$

where x is a point of X and ψ is a local C^∞ map defined in a neighborhood of x. Two such pairs (x,ψ), (x',ψ') are *equivalent* if the following conditions are satisfied:

$$x = x'$$

$$\psi(x) = \psi'(x)$$

ψ and ψ' agree to first order at x, i.e., $\qquad (4.2)$

$$\psi_*(v) = \psi'_*(v)$$

for all $v \in X_x$

$J^1(X,Y)$, the space of *Ehresmann one-jets of maps from* X *to* Y, is the quotient of the pairs of form (4.1) under this equivalence relation.

Let

$L(T(X),T(Y))$ = set of ordered triples (x,y,γ), where

$$x \in X, \quad y \in Y, \qquad (4.3)$$

γ is a linear map: $X_x \to Y_y$.

Assign to $(x,y,\gamma) \in L(T(X),T(Y))$ the point $(x,y) \in X \times Y$. This assignment makes $L(T(X),T(Y))$ a vector bundle over $X \times Y$: The fiber is a real vector space of dimension nm, where $n = \dim X$, $m = \dim Y$. (Intrinsically, the fiber is the vector space $L(X_x, Y_y)$ of linear maps: $X_x \to Y_y$.)

Given a pair (x,ψ) of the form (4.1), we can assign an element of $L(T(X),T(Y))$, namely, the triple

$$(x, \psi(x), \psi_*) \: . \qquad (4.4)$$

This assignment is constant on the equivalence classes which define $J^1(X,Y)$, hence defines a map

$$J^1(X,Y) \to L(T(X),T(Y)) \ .$$

Theorem 4.1. This assignment identifies $J^1(X,Y)$ with the vector bundle $L(T(X),T(Y))$.

The proof is readily made by introducing local coordinates for X and Y.

Theorem 4.2. The dual vector bundle to $L(T(X),T(Y))$ can be naturally identified with $L(T(Y),T(X))$.

Proof. Let (x,y,γ) and (y,x,γ') be elements of $L(T(X),T(Y))$ and $L(T(Y),T(X))$,

$$x \in X, \quad y \in Y$$

$$\gamma \in L(X_x, Y_y)$$

$$\gamma' \in L(Y_y, X_x) \ .$$

Define

$$< (x,y,\gamma), (y,x,\gamma') > \ = \ \text{trace } (\gamma'\gamma) \ . \qquad (4.5)$$

It is readily verified that this formula defines one bundle as the dual of the other. Q.E.D.

Let us now use local coordinates (x^μ) for X, (y^a) for Y, with the range of indices and notation used in Sections 2 and 3, to define coordinates for $L(T(Y),T(X))$, hence for the dual of $J^1(X,Y)$.

Let

$$\partial_\mu \ , \quad \partial_a$$

be the basic vector fields on X and Y defined by these coordinates. For $(x,y,\gamma \in X \times Y \times L(Y_y, X_x))$ define the matrix π_a^μ such that:

$$\gamma(\partial_a(y)) \ = \ \pi_a^\mu \partial_\mu(x) \ . \qquad (4.6)$$

Thus, assigning to

$$(x,y,\gamma) \in L(T(X),T(Y))$$

the numbers

$$(x^\mu(x), \phi^a(y), \pi_a^\mu) \tag{4.7}$$

defines local coordinates for

$$M \equiv L(T(Y), T(X)) \ . \tag{4.8}$$

Theorem 4.3. Formula (4.7) defines functions $(x^\mu, \phi^a, \pi_a^\mu)$ for M, which defines local coordinates.

Proof. Routine verification, left to the reader.

We can now go back to Section 3 and identify *intrinsically* certain differential forms on M.

Theorem 4.4. The n-form

$$\eta = \pi_a^\mu \, d\phi^a \wedge dx_\mu$$

on M is intrinsically defined, independently of the coordinate system.

Proof. This follows from (4.6) and linear algebra.

We can now see how, in the following case,

$$n = \dim X = 1 \tag{4.10}$$

i.e., the "particle mechanics" case, η reduces to the usual *contact* form. There is but one coordinate "x^0", hence,

$$dx_\mu = \frac{\partial}{\partial x^0} \, \lrcorner \, dx^0$$

$$= 1$$

is a zero-th degree form. Set:

$$\pi_a^0 = p_a \ , \tag{4.11}$$

and (4.9) takes the following form:

$$\eta = p_a \, dy^a \ , \tag{4.12}$$

which is the familiar formula for the contact form on $T^d(Y)$, cotangent bundle to Y. $L(T(Y), T(X))$ is, in this case, isomorphic to

$$T^d(Y) \times X \ .$$

Hamilton-Volterra and Cojet Bundles

The form θ given by formula (3.6) can be written as follows:

$$\theta = -\eta + H \, dx^0 , \qquad (4.13)$$

which, if "x^0" is changed to "t", becomes the familiar Poincare-Cartan form, in more familiar notation

$$\theta = -p_a \, dq^a + H \, dt . \qquad (4.14)$$

So far, H is an arbitrary real-valued function on $L(T(Y),T(X))$. In the particle mechanics case, H is the "total energy". In field theories the physical interpretation of it is more obscure. I now want to describe it as the trace of another tensor, which I will call the *Volterra tensor*.

5. VOLTERRA TENSORS

Continue with the notation of previous sections, which we recapitulate. X and Y are finite dimensional manifolds. The one-jet space $J^1(X,Y)$ may be identified with the vector bundle

$$L(T(X),T(Y))$$

of linear bundle maps of the tangent bundle $T(X)$ into $T(Y)$. Let M be the dual vector bundle, i.e., M has a submersion map

$$\pi: M \to X \times Y ,$$

and the fiber of π above the point (x,y) of $X \times Y$ is the space

$$L(Y_y, X_x)$$

of linear maps: $Y_y \to X_x$.

If x^μ and ϕ^a are coordinates for X and Y, then denote by

$$(x^\mu, \phi^a, \phi^a_\mu)$$

the natural coordinates for $J^1(X,Y)$. The (ϕ^a_μ) are linear functions on the linear vector space fibers of $J^1(X,Y) \equiv L(T(X),T(Y))$. Let

$$(x_\mu, \phi^a, \pi^\mu_a)$$

be the coordinate system for M, with the following property:

(π^μ_a) are the dual linear functions to ϕ^a_μ, i.e., the inner product on the fibers is the map

$$(x,y,\sigma), (x,y,\sigma^d) \to \phi^a_\mu(\sigma)\pi^\mu_a(\sigma^d) \qquad (5.1)$$

$$\equiv \langle (x,y,\sigma), (x,y,\sigma^d) \rangle$$

Let $H: M \to R$ be a real-valued, C^∞ function and let

$$\theta = \pi^\mu_a \, d\phi^a \wedge dx_\mu - H \, dx \qquad (5.2)$$

be the Cartan form defined in previous sections. (Notice that this is the negative of the form θ defined in Section 3. We will work with this sign convention in order to conform with the conventions in the analytical mechanics books.)

Suppose now that

$$T^\nu_\mu(x,\phi,\pi)$$

is a matrix of functions (that we call the *Volterra tensor*) such that:

$$T^\mu_\mu = H \, . \qquad (5.3)$$

Then,

$$H \, dx = dx^\mu \wedge (T^\nu_\mu \, dx) \, . \qquad (5.4)$$

Substitute (5.4) into (5.2):

$$\theta = \pi^\mu_a \, d\phi^a \wedge dx_\mu - dx^\nu \wedge T^\mu_\nu \, dx_\mu$$

$$= (\pi^\mu_a d\phi^a - T^\mu_\nu dx^\nu) \wedge dx_\mu \, . \qquad (5.5)$$

Hence,

$$d\theta = (d\pi^\mu_a \wedge d\phi^a - dT^\mu_\nu \wedge dx^\nu) \wedge dx_\mu \, . \qquad (5.6)$$

Now, set:

$$\omega^\mu = d\pi^\mu_a \wedge d\phi^a - dT^\mu_\nu \wedge dx^\nu \, . \qquad (5.7)$$

Then,

$$d\theta = \omega^\mu \wedge dx_\mu .\tag{5.8}$$

Thus, in order to get decomposition of the form (5.7), we can suppose that ω^μ is written in the following form:

$$\omega^\mu = \alpha_a^\mu \wedge \beta^a ,\tag{5.9}$$

where (α_a^μ, β^a) are a new set of forms (which will depend on T_μ^ν). Then,

$$d\theta = \alpha_a^\mu \wedge \beta^a \wedge dx_\mu$$

$$= \pm (\alpha_a^\mu dx_\mu) \wedge \beta^a .\tag{5.10}$$

This, together with the work of Section 3, suggests that the extremals (i.e., the solution of the Hamilton-Volterra equations) will be integral submanifolds of the exterior differential system:

$$\beta^a = 0 = \alpha_a^\mu \wedge dx_\mu .\tag{5.11}$$

As an example, let us turn to the harmonic map/σ-model examples.

6. THE NATURAL CLASS OF FIELD THEORIES CONTAINING THE HARMONIC MAPS OF EELS AND SAMPSON, THE σ-MODEL OF THE ELEMENTARY PARTICLE PHYSICISTS, AND GENERALIZING THE "NEWTONIAN" MECHANICAL MODELS

Continue with the notation of Section 5. Suppose that the Volterra tensor (T_μ^ν) is given by the following formula

$$T_\mu^\nu = \tfrac{1}{2} g_{\mu\mu'} g^{ab} \pi_a^{\mu'} \pi_b^\nu + \tfrac{1}{n} V(\phi) \delta_\mu^\nu ,\tag{6.1}$$

where:

$$g_{ab}\, d\phi^a d\phi^b \quad \text{is a Riemannian metric on } Y \tag{6.2}$$

$$(g^{ab}) \text{ is the inverse matrix to } (g_{ab}) \tag{6.3}$$

$$g_{\mu\mu'}\, dx^\mu dx^{\mu'} \text{ is a Riemannian metric on } X \tag{6.4}$$

V is a function on Y.

Let us calculate the Volterra equations:

$$H = \frac{1}{2} g_{\mu\mu'} g^{ab} \pi_a^{\mu'} \pi_b^{\mu} + V(\phi) \tag{6.6}$$

$$H_a = \frac{1}{2} g_{\mu\nu} g_a^{a'b} \pi_{a'}^{\mu} \pi_b^{\nu} + V_a \tag{6.7}$$

(Subscripts denote partial derivatives.)

$$H_\mu^a \equiv \frac{\partial H}{\partial \pi_a^\mu} = g_{\mu\nu} g^{ab} \pi_b^\nu \tag{6.8}$$

Thus, the Hamilton-Volterra equations take the following form:

$$\frac{\partial \phi^a}{\partial x^\mu} = H_\mu^a$$

$$= g_{\mu\nu} g^{ab} \pi_b^\nu$$

$$\frac{\partial \pi_a^\mu}{\partial x^\mu} = -H_a$$

$$= -\frac{1}{2} g_{\mu\nu}(x) \frac{\partial g^{a'b}}{\partial \phi^a} \pi_{a'}^{\mu} \pi_b^{\nu} - \frac{\partial V}{\partial \phi^a}$$

These equations can be readily put into a coordinate-free form using either *orthonormal frames* for the Riemannian metrics on X and Y, or, alternately, using the covariant derivative operations involving these metrics. In Ref. 4, I have computed *currents*, and the quantum-mechanical *commutation relations* for these currents, in terms of the geometric operations associated with these Riemannian metrics.

References

1. R. Hermann, The geometric foundations of the integrability property and differential systems, Part I, Lie's "Function Groups", *J. Math. Phys.* **24**, 2422-2432 (1983); Part II, Lax representations of dynamical systems and analytical mechanics on affinely connected manifolds, to appear, *J. Math. Phys.*

2. V. Volterra, *Collected Works*, Vol. I, 1890.

3. R. Hermann, *Lie Algebras and Quantum Mechanics*, W.A. Benjamin, Reading MA, 1970.

4. R. Hermann, Current algebras, the Sugawara model, and differential geometry, *J. Math. Phys.* 11, 1825-1829 (1970); Geometric formula for current-algebra commutation relations, *Phys. Rev.* 177, 2449 (1969); Quantum field theories with degenerate Lagrangians, *Phys. Rev.* 177, 2453 (1969); Dynamical systems defined on infinite dimensional Lie algebras of "current algebra" or "Kac-Moody" type, *AIP Conf. Proc.*, No. 88, M. Tabor and Y. Treve (eds.), 1982.

5. R. Hermann, Infinite dimensional Lie algebras and current algebras, *Proc. 1969 Battelle Rencontre*, Springer-Verlag, Berlin, 1970.

6. M. Gell-Mann and M. Levy, *Nuovo Cimento* 16, 705 (1960); T.D. Lee, S. Weinberg, and B. Zumino, *Phys. Rev. Lett.* 18, 1029 (1967); K. Bardacki, Y. Frishman, and M.B. Halpern, *Phys. Rev.* 170, 1353 (1968); H. Sugawara, *Phys. Rev.* 170, 1659 (1968); K. Bardacki and M.B. Halpern, *Phys. Rev.* 172, 1542 (1968); H. Sugawara and M. Yoshimura, *Phys. Rev.* 173, 1419 (1968); C. Sommerfield, *Phys. Rev.* 176, 2019 (1968); L. Dolan, *Phys. Rev. Lett.* 47, 1371 (1981).

7. S. Lie, *Transformationsgruppen*, Vol. II, Chelsea Publishing Co., New York.

8. C. Ehresmann: a) Sur les structures infinitesimales regulieres, *Congres Intern. Math. Amsterdam* 1, 479-480 (1954). b) Connexions infinitesimales, *Colloque Top. Alg. Bruxelles*, 29-55 (1950); c) Structures infinitesimals et pseudogroupes de Lie, *Colloq. Intern C.N.R.S. Geom. Diff. Strasbourg*, 97-110 (1953). d) *Compte-rendus Acad. Sc. Paris* 240 (1954); 241 (1955), 397 and 1755; 246 (1958), 360. e) Connexions d'ordre superieur, *Atti 5 Congr. dell'Unione Mat. Italiana* 1955, Ed. Cremonese, Roma, 326-328 (1956). f) Categories topologiques et categories differentiables, *Colloq. Geom. Diff. Globale Bruxelles*, C.B.R.M., 137-150 (1958). g) Groupoides differenciales, *Revista Un. Mat. Argentina* XIX, Buenos-Aires, 48 (1960).

9. E. Cartan, *Lecons sur les Invariants Integraux*, Herman, Paris, 1922.

10. E. Cartan, *Les Systemes Differentiels Exterieurs et leurs Applications Geometriques*, Herman, Paris, 1945.

11. R. Hermann, *Geometry, Physics and Systems*, Marcel Dekker, New York, 1973.

12. J. Kijowski and W. Tulczyjew, *Lecture Notes in Physics*, No. 109, Springer-Verlag.

12. R. Hermann, *Differential Geometry and the Calculus of Variations*, 2nd Ed., Math Sci Press, 1977.

13. H. Goldschmidt and S. Sternberg, *Ann. Inst. Fourier (Grenoble)* 231, 203 (1973).

14. W. Shadwick, *Letters in Math. Phys.* 6, 409-416 (1982); 5, 137 (1981); 4, 241 (1980).

15. Th. de Donder, *Theorie Invariantive du Calcul des Variations*, Gauthier-Villars, Paris, 1935.

16. P. Dedecker, *C.R. Acad. Sci. Paris* 288, 827 (1879).

RAYMOND H. FOGLER LIBRARY
DATE DUE

BOOKS ARE SUBJECT TO
RECALL AFTER TWO WEEKS

MAY 1 4 1987